T0212229

Fundamentals of
Biomedical Transport Processes

Fundamentals of Biomedical Transport Processes
Gerald E. Miller

ISBN: 978-3-031-00517-6   paperback
ISBN: 978-3-031-01645-5   ebook

DOI 10.1007/978-3-031-01645-5

A Publication in the Springer series
SYNTHESIS LECTURES ON BIOMEDICAL ENGINEERING

Lecture #37
Series Editor: John D. Enderle, *University of Connecticut*
Series ISSN
Synthesis Lectures on Biomedical Engineering
Print 1930-0328   Electronic 1930-0336

Synthesis Lectures on Biomedical Engineering

Editor
John D. Enderle, *University of Connecticut*

Lectures in Biomedical Engineering will be comprised of 75- to 150-page publications on advanced and state-of-the-art topics that spans the field of biomedical engineering, from the atom and molecule to large diagnostic equipment. Each lecture covers, for that topic, the fundamental principles in a unified manner, develops underlying concepts needed for sequential material, and progresses to more advanced topics. Computer software and multimedia, when appropriate and available, is included for simulation, computation, visualization and design. The authors selected to write the lectures are leading experts on the subject who have extensive background in theory, application and design.

The series is designed to meet the demands of the 21st century technology and the rapid advancements in the all-encompassing field of biomedical engineering that includes biochemical, biomaterials, biomechanics, bioinstrumentation, physiological modeling, biosignal processing, bioinformatics, biocomplexity, medical and molecular imaging, rehabilitation engineering, biomimetic nano-electrokinetics, biosensors, biotechnology, clinical engineering, biomedical devices, drug discovery and delivery systems, tissue engineering, proteomics, functional genomics, molecular and cellular engineering.

Quantitative Neurophysiology
Joseph V. Tranquillo
2008

Tremor: From Pathogenesis to Treatment
Giuliana Grimaldi and Mario Manto
2008

Introduction to Continuum Biomechanics
Kyriacos A. Athanasiou and Roman M. Natoli
2008

The Effects of Hypergravity and Microgravity on Biomedical Experiments
Thais Russomano, Gustavo Dalmarco, and Felipe Prehn Falcão
2008

A Biosystems Approach to Industrial Patient Monitoring and Diagnostic Devices
Gail Baura
2008

Multimodal Imaging in Neurology: Special Focus on MRI Applications and MEG
Hans-Peter Müller and Jan Kassubek
2007

Estimation of Cortical Connectivity in Humans: Advanced Signal Processing Techniques
Laura Astolfi and Fabio Babiloni
2007

Brain–Machine Interface Engineering
Justin C. Sanchez and José C. Principe
2007

Introduction to Statistics for Biomedical Engineers
Kristina M. Ropella
2007

Capstone Design Courses: Producing Industry-Ready Biomedical Engineers
Jay R. Goldberg
2007

BioNanotechnology
Elisabeth S. Papazoglou and Aravind Parthasarathy
2007

Bioinstrumentation
John D. Enderle
2006

Fundamentals of Biomedical Transport Processes

Gerald E. Miller
Virginia Commonwealth University

SYNTHESIS LECTURES ON BIOMEDICAL ENGINEERING #37

ABSTRACT

Transport processes represents important life sustaining elements in all humans. These include mass transfer processes, including gas exchange in the lungs, transport across capillaries and alveoli, transport across the kidneys, and transport across cell membranes. These mass transfer processes affect how oxygen and carbon dioxide are exchanged in your bloodstream, how metabolic waste products are removed from your blood, how nutrients are transported to tissues, and how all cells function throughout the body. A discussion of kidney dialysis and gas exchange mechanisms is included.

Another element in biomedical transport processes is that of momentum transport and fluid flow. This describes how blood is propelled from the heart and throughout the cardiovascular system, how blood elements affect the body, including gas exchange, infection control, clotting of blood, and blood flow resistance, which affects cardiac work. A discussion of the measurement of the blood resistance to flow (viscosity), blood flow, and pressure is also included.

A third element in transport processes in the human body is that of heat transfer, including heat transfer inside the body towards the periphery as well as heat transfer from the body to the environment. A discussion of temperature measurements and body protection in extreme heat conditions is also included.

KEYWORDS

biomedical mass transfer, biomedical fluid mechanics, biomedical heat transport, dialysis, biomedical flow, and pressure measurement

This book is dedicated to my wife, Mary,
who has come through many life trials and
has served as a source of inspiration.

Contents

CHAPTER 1

Biomedical Mass Transport

Mass transport within the human body is a vital process that affects how the lungs function in transferring air and its components to the bloodstream, how the capillaries function in transferring nutrients and gases to surrounding body tissues, and how the kidneys function in transferring metabolic waste products and excess water from the blood into the urine. Mass transfer processes also occur in artificial devices such as artificial kidneys (dialysis), artificial ventilators, and respirators. Mass transfer in the body also affects transport across cell membranes, which controls processes in millions of cells affecting every area of the body. The majority of mass transfer occurs across small membranes of thin surfaces in order to shorten the distance over which substances must travel from point A to point B. This is true of cells in the body, which are quite thin as well as artificial devices whose components are manufactured to be very thin.

1.1   ANALYSIS OF RESPIRATION AND GAS TRANSPORT

The human lungs control gas exchange from our environment into the bloodstream by means of pressure and concentration gradients. When breathing, air enters the lungs through a large entrance, the trachea, and eventually branches into smaller and smaller segments until reaching the smallest elements of the lungs, the alveoli. Each of these thin elements is in close proximity to blood in pulmonary capillaries, which are the smallest and thinnest of the blood vessels. With each of the alveoli in close proximity to a pulmonary capillary, the distance for gas exchange is very short, which thus shortens the time by which complete gas exchange occurs. A diagram of the lungs and its branching system is shown in Figure 1.1.

The two human lungs contain approximately 300 to 500 million alveoli having a total surface area of about 75 m^2 in adults, the size of a tennis court. The branching of the airways into the alveoli represents millions of tiny sacs, which not only represents thinner membranes to speed gas exchange but also more surface area to speed gas exchange. The alveolar network is shown in Figure 1.2.

The lungs may be separated (for the purposes of mass transfer/gas exchange) as a dead space and an alveolar space. These dead spaces, consisting of the trachea, the bronchioles and the bronchi, are all large segments of the airway where there is air flow but no gas exchange with the bloodstream. The alveolar space is where the actual gas exchange occurs. However, both zones heat the inspired air to body temperature (37 degrees C) as well as humidify the air. Thus, expired air is heated to body temperature and is usually fully saturated with water vapor. In fact, it is possible to lose up to a half pound per day, merely from losing water from the body via respiration.

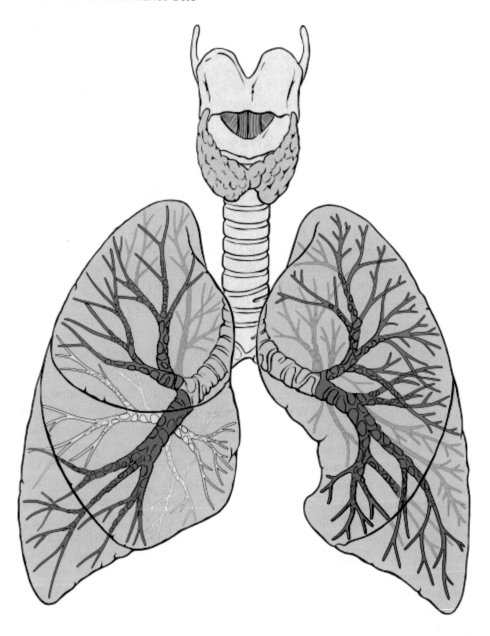

Figure 1.1: Branching of the airway network within the human lungs from the trachea ending in alveoli (Patrick J. Lynch; illustrator; C. Carl Jaffe; MD; cardiologist Yale University Center for Advanced Instructional Media Medical Illustrations by Patrick Lynch, generated for multimedia teaching projects by the Yale University School of Medicine, Center for Advanced Instructional Media, 1987-2000).

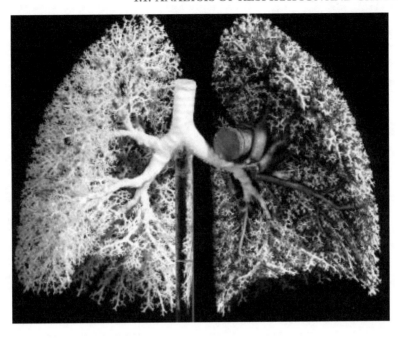

Figure 1.2: A cast of the human lungs showing the alveolar sacs (Anatomical Institute, Bern).

The amount of air which one inhales and exhales (without exertion) is called the *tidal volume*, which is 500 ml (about a fluid pint) per breath. The typical breathing rate is 12 breaths per minute at rest. Ambient air is approximately 79% nitrogen and 21% oxygen on a dry basis (not including any water vapor/humidity in the air). With a single breath, oxygen is transferred from the air to the pulmonary blood, and carbon dioxide is transferred from the blood to the air, across the alveolar membranes. Nitrogen is not transferred under normal conditions. Figure 1.3 depicts the gas exchange between alveoli and pulmonary capillaries.

The resulting single breath gas exchange is summarized in the Table 1.1 shown below. The gases (nitrogen, oxygen, carbon dioxide) are shown with "wet" percentages, which incorporates the water vapor in the air.

For the purposes of gas exchange, the components of the inspired air are described by means of their partial pressures. This is the fractional amount of total gas pressure due to the substance being measured. For example, at sea level, the total atmospheric pressure is 760 mm Hg. The amount of pressure that is due to oxygen is approximately $0.21 \times 760 = 160$ mm Hg. We would say that the partial pressure of oxygen at sea level in dry air (no vapor in the air) is 160 mm Hg. The partial pressure of carbon dioxide in dry air at sea level is $0.03 \times 760 = 22.8$ mm Hg. The maximum amount of water vapor in the air varies with temperature and relative humidity. At body temperature (37

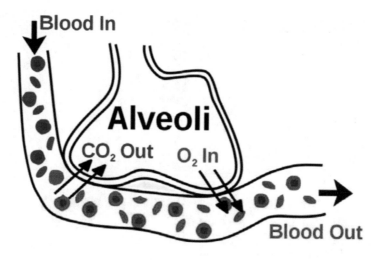

Figure 1.3: Gas exchange between alveoli and pulmonary capillaries. This event occurs across all of the millions of alveoli and associated capillaries. (Wikipedia Respiratory Gas Exchange).

Table 1.1: Single breath analysis for components of air. Note that not all of the oxygen in the air is taken up by the lungs and that the expired gases from the lungs are heated to body temperature and saturated with water vapor – thus the higher water vapor content of the expired air. There is trace amount of carbon dioxide in the ambient air but a higher percentage in the expired air. Thus, oxygen replaces carbon dioxide in the pulmonary bloodstream.

Air Component	Atmospheric Air %	Expired Air %
Nitrogen and Inert Gases	78.62	74.9
Oxygen	20.85	15.3
Carbon Dioxide	0.03	3.6
Water Vapor	0.5	6.2
Total	100	100

degrees centigrade), air can be saturated up to 47 mm Hg of water vapor pressure. Therefore, in the lungs, where air is totally water-saturated, the partial pressure of water vapor would be 47 mm Hg.

Thus, in terms of partial pressures, and with the ambient humidity equal to a partial pressure of 10 mm Hg (as an example), Table 1.1 can be rewritten as shown below. The partial pressure of water vapor is 10 mm Hg and the partial volume would = $(10/760) \times 500$ ml tidal volume = 6.5 ml. The dry gases (nitrogen and oxygen) represent the remainder of the tidal volume, or $500 - 6.5 =$

493.5. Thus, the nitrogen partial volume would be 493.5 × 79% = 389.9 ml, and the partial volume of oxygen would be 493.5 × 21% = 103.6 ml.

The nitrogen is not exchanged in the lungs, and thus its partial volume does not change. However, the oxygen is taken up by the bloodstream, and the carbon dioxide is released by the bloodstream to be exhaled from the lungs. The partial volumes of the exhaled gases are based upon the metabolic load of the body. At rest, these are 284 ml/min of oxygen (at body temperature) and 226 ml/min for carbon dioxide. This equals 23.67 ml/breath for oxygen (284/12 breaths per minute) and 18.83 ml/breath for carbon dioxide (226/12). The water vapor during exhalation equals the partial pressure of water vapor in the body (47 mm Hg), compared to the dry gas pressure (760−47 = 713 mm Hg) times the dry gas expired volume. Thus, the water vapor attaches to the dry gas during expiration.

Table 1.2: Single breath analysis based upon partial volumes.		
Air Component	Inspired Gases Partial Vol.	Expired Gases Partial Vol.
Water Vapor	6.5	$6.5 + (47/713) \times 488.63 = 38.7$ ml
Dry Gas	493.5	$389.9 + 79.9 + 18.83 = 488.63$
Nitrogen	389.9	389.9
Oxygen	103.6	$103.6 - 23.67 = 79.9$
Carbon Dioxide	0.0	$0.0 + 18.83 = 18.83$

The above analysis does not consider expansion and contraction due to temperature differences, which do, in fact, exist. This is beyond the scope of this discussion, but it is available from several other sources.

In summary, gas exchange between the lungs and the pulmonary circulation occurs at the smallest elements of each, which are also in extremely close proximity to each other. With so many alveoli, the surface area for gas exchange is very large, which promotes faster mass transfer. The mass transfer of each of the components of air depends on the partial pressure of each as well as the metabolic load of the body, along with the diffusion distance between alveoli and pulmonary capillaries. Oxygen is taken up by the bloodstream, and carbon dioxide is given off to the lungs from the bloodstream. The exhaled gases are heated to body temperature and are fully saturated with water vapor. Thus, when you place your hand over your mouth as you exhale, the air feels warm and moist.

1.2   MEMBRANES, PORES AND DIFFUSION

The gas exchange between the alveoli and the pulmonary capillaries is affected by the partial pressures of each gas, the distance over which mass transfer occurs, and the surface area for mass transfer. We can think of mass transfer as occurring across a membrane in a steady state fashion. This can be described by the one-dimensional, steady state form of *Fick's Law*:

$$\text{Mass exchange rate} = DA \, dC/dx \, .$$

where D is the *diffusivity*, A is the surface area for mass transfer, and dC/dx is the concentration gradient for mass transfer.

For gases, the concentration gradient is the difference in the *partial pressure*. For a liquid or for substances dissolved in a liquid (such as in blood or extracellular fluid), the concentration gradient can be approximated by DC/DX, the difference in concentration over a specified distance. The *diffusivity* (D) is a parameter which describes the relative ease by which a substance moves throughout the "medium" – the fluid through which the substance moves. As an example, how sodium ions move through extracellular fluid. The diffusivity is thus affected by both the medium and the substance to be transferred – it is a material property of both.

For gases, mass transfer occurs across the entire membrane, as gases are lipid soluble, with membranes consisting of a layer of lipids and proteins. However, when the exchange is of a liquid or a substance dissolved in a liquid (such as an ion), then the mass transfer must occur though pores in the membrane. That is because most liquids are not lipid soluble. When substances do indeed travel across pores, then there is selective mass transfer due to the size of the pore as compared to the size of the substance which might transfer. This selective mass transfer by means of the relative sizes is called *filtration*. Fick's Law is modified to account for such selective mass transfer by adding a permeability term (P), which relates the area of the pores as compared to the total area of the membrane. At times, the area for mass transfer (A) is instead described as Ap rather than as the product of $A \times P$. An example of filtration is seen in Figure 1.4.

In general, the process by which mass transfer occurs via a concentration gradient is called *diffusion*. Diffusion is the macroscopic result of random thermal motion on a microscopic scale. For example, in Figure 1.5, oxygen and nitrogen molecules move in random directions, with kinetic energy on the order of kT, where k is the thermal conductivity and T is the absolute temperature. If there are more oxygen molecules on the left side of the plane A-A than on the right, more molecules will cross to the right than to the left: there will be a net movement even though the motion of each individual molecule is completely random. Diffusion in an open environment such as within extracellular fluid, and not constrained by a membrane for mass transfer, is shown in Figure 1.5.

Again, the net movement of any substance (within a mixture of substances as shown above) across line A-A is a function of the concentration of that substance on each side of the line. It is certainly possible that the net movement of one substance is in the opposite direction of another – each based on its own concentration gradient. This is also true for mass transfer across pores. The concentration gradients control the amount and the direction of mass transfer for each substance. When a substance, such as an ion, is in solution (such as within extracellular fluid), then the substance is known as a *solute* and the fluid as a *solvent*. Mass transfer of water is known as *osmosis*.

Further, the relative size of the solute as compared to the pore size, may affect the potential mass transfer by either (1) completing restricting the movement (if the species is larger than the pore size) or (2) merely restrict it because it may be nearly equal to the pore size. In such cases, Fick's

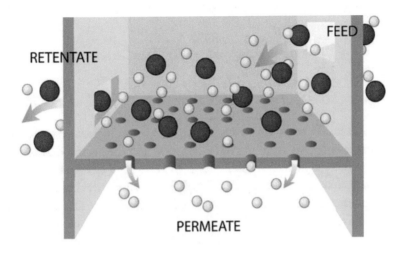

Figure 1.4: Filtration - selective mass transfer by means of pore size and size of substances compared to pore size (Wikipedia).

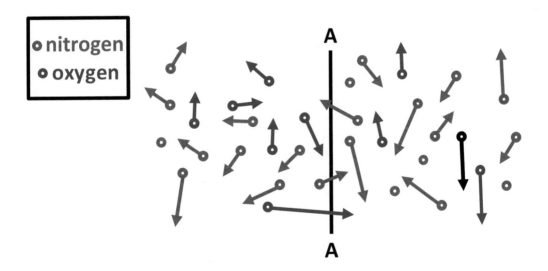

Figure 1.5: Random movement of particles resulting in a net movement across a dividing line based upon relative concentrations on both sides of the line (Time Domain CVD).

law can be modified to account not only for the limited pore area (via the permeability), but also via the pore to species size, which is given by the *restrictive diffusivity* (D_R). Thus, Fick's law could be rewritten as:

$$\text{Mass exchange rate} = D_R \, A_P \, dC/dx \,.$$

1.3   MASS TRANSPORT IN SYSTEMIC CAPILLARIES

Capillaries, the smallest and most numerous of the blood vessels, form the connection between the vessels that carry blood away from the heart (arteries) and the vessels that return blood to the heart (veins). The primary function of capillaries is the exchange of materials between the blood and tissue cells. Gas exchange occurs across capillary walls in a manner similar to that which occurs across alveoli and pulmonary capillaries. A partial pressure gradient exists for each gas (primarily oxygen and carbon dioxide) which controls the gas exchange process. As with pulmonary capillaries, the gas transfer across systemic capillaries occurs across the entire membrane wall, as the gases are lipid soluble.

However, unlike the pulmonary capillaries, mass transfer across systemic capillaries also includes liquids and ions (dissolved or free floating within the liquid) which travel across pores in the capillary walls. In addition to forming the connection between the arteries and veins, systemic capillaries have a vital role in the exchange of gases, nutrients, and metabolic waste products between the blood and the tissue cells. Substances pass through the capillary wall by diffusion, filtration, and osmosis. Oxygen and carbon dioxide move across the capillary wall by diffusion regulated by the partial pressure differences. Fluid movement across a capillary wall via the pores is determined by a combination of hydrostatic and osmotic pressure. The net result of the capillary microcirculation created by hydrostatic and osmotic pressure is that substances leave the blood at one end of the capillary and return at the other end, as the hydrostatic pressure drops along the length of the capillary, and thus the pressure difference (hydrostatic – osmotic) is different from the beginning of the capillary (arteriole side) to the end (venule side).

Thus, the driving mechanism for this mass transfer is twofold. The pressure (called the *hydrostatic pressure*) is a pushing pressure. Opposite in direction is the *osmotic pressure*, which is a pulling pressure, and which results from the concentration difference of substances that cannot fit through the pores. These are the non-diffusible components. These two pressure differences, acting across the tube-like pores of the capillary, are shown in Figure 1.6.

The capillaries form a network of minute vessels between arterioles and venules so as to maximize mass transfer by (a) shortening the distance for mass transfer and (b) maximizing the overall surface area for mass transfer. This is shown in Figure 1.7.

The anatomy of capillaries is well suited to the task of efficient exchange. Capillary walls are composed of a single layer of endothelial cells. The thin nature of the capillary wall facilitates efficient diffusion of oxygen and carbon dioxide as well as containing short pore lengths to facilitate the bulk motion of liquids and dissolved ions. This is seen in Figure 1.8.

Capillary Microcirculation

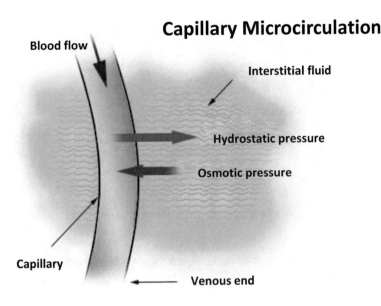

Blood flow

Interstitial fluid

Hydrostatic pressure

Osmotic pressure

Capillary

Venous end

Figure 1.6: Mass transfer across systemic capillaries.

As was stated above, the hydrostatic pressure within a capillary is higher on the arteriole end of the capillary and lower on the venule end. This is due to the normal pressure gradient along a pipe or tube that produces the axial (down the vessel) flow rate. In fact, a pressure drop is required in order to propel the blood downstream. As a result, the higher hydrostatic pressure near the arteriole end is greater than the osmotic pressure, and the net pressure is outward – pushing fluid out of the capillary. As the hydrostatic pressure drops along the length of the capillary, the pressure difference (hydrostatic – osmotic) is reduced, and eventually reverses as blood reaches the venule end. Thus, there is a net inflow of fluid at the venule end (into the capillary).

In addition, arterioles are known as the resistance vessels in the circulatory system; in that, they can constrict or dilate in response to neural and/or hormonal feedback, given their large smooth muscle composition. Arterioles constrict in response to a decrease in arterial pressure and dilate in response to a rise in arterial pressure. When arterioles constrict, they produce an added resistance to blood flow, which drops the hydrostatic pressure downstream. This is because the pressure gradient is proportional to the blood vessel resistance. When the downstream pressure at the end of arterioles is reduced, this corresponds to the inlet pressure to capillaries. Therefore, with arteriolar constriction, the hydrostatic pressure is reduced all along the capillary, which results in less pushing pressure out of the capillary. This produces a net increase in fluid entering the capillary, and eventually leaving the capillary via blood flow into the venules. With the increase in fluid level, the blood volume increases,

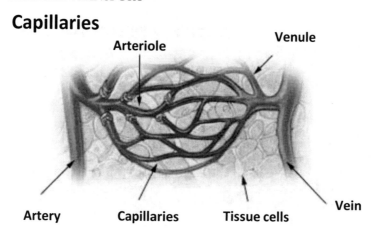

Capillaries

Figure 1.7: Capillary network allowing close proximity to tissues, which shortens the distance for mass transfer.

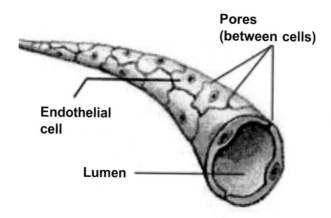

Figure 1.8: Thin wall of a systemic capillary which facilitates mass transfer by creating a short distance for transport.

which results in greater venous return of blood into the heart and a resulting increase in cardiac output and arterial pressure. This is shown in the block diagram of Figure 1.9.

With dilation of arterioles, the opposite effect is created. The resistance of the arteriole is reduced, which produces a smaller pressure gradient across the arteriole. This then produces a rise in the inlet hydrostatic pressure to capillaries which results in a greater pushing pressure for fluid moving out of the capillary into the tissues and extracellular fluid. With a net increase in fluid leaving the capillaries, there is a reduction in blood volume and in venous return. This results in a reduction

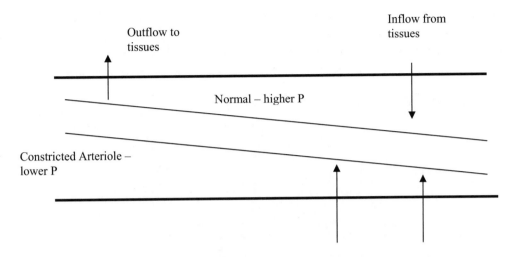

Figure 1.9: Hydrostatic pressure gradient inside a capillary for a normal condition and for a constricted arteriole. The hydrostatic pressure in reduced all along the capillary from the resulting reduction at the inlet. This produces a greater inflow of fluid into the capillary.

in cardiac output and in arterial pressure. Table 1.3 summarizes the cause and effect of changes in arteriole diameter.

Table 1.3: Relationship between arteriole diameter, capillary hydrostatic pressure, trans-capillary fluid flow, and arterial pressure.

Initial Arterial Pressure	Arteriole Diameter Change	Capillary Hydrostatic Effect	Net Fluid	Venous Return	Cardiac/Arterial Output/Pressure
Low	constrict	reduced	into capillary	increase	increase
High	dilate	increased	out of capillary	decrease	decrease

When there is an increase in fluid leaving the capillary as a result of arteriole dilation, then the extracellular fluid pressure rises. The increase in fluid volume will eventually cause edema, a buildup of fluid which produces swelling and potential damage to cells and tissues. To alleviate this condition, there is a network of *lymphatic vessels* that parallels the circulatory system. The peripheral lymphatic vessels are similar in size to capillaries and are networked alongside the capillaries. Their function is to collect excess fluid from extracellular spaces and transport it along the lymphatic vessel network of larger and larger vessels until the lymph reaches the vena cava where it is returned to the bloodstream. The lymph flow is normally quite small compared with blood flow. Even the terminal lymph flow is quite small (1 ml/hr), compared with capillary axial blood flow (1 ml/sec). The terminal lymphatic vessels located in close proximity to systemic capillaries are seen in Figure 1.10.

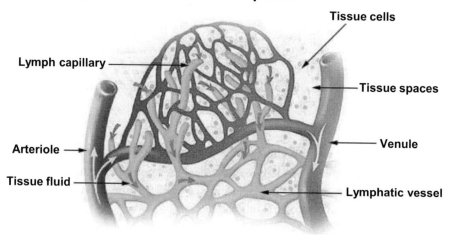

Figure 1.10: Lymphatic vessels embedded within the systemic capillary bed (Wikipedia).

Propulsion of lymph through the larger vessels is produced by contraction of the walls of the vessels in a fashion similar to that provided for flow through the human intestines and somewhat similar to the action of muscles on veins in the cardiovascular system. Lymph vessels, like the venous system, have valves within the vessels to limit backflow and to promote forward flow along the circulation. This is in contrast with flow through the bloodstream, which is produced by a driving hydrostatic pressure gradient from the aorta through arteries, arterioles, capillaries, venules, and veins. The lymph flow originates in the terminal vessels and therefore has no embedded driving pressure gradient. The contraction of the lymph vessels in concert with the added fluid volume (from all terminal lymph vessels joining together within the network) produces the required lymph flow headed to the vena cava.

1.4   MASS TRANSPORT IN THE KIDNEYS AND DIALYSIS

As with the alveoli in the lungs, mass transfer in the kidneys is also processed via a large array of extremely small elements. Each of these elements is a *nephron*. Although nephrons and alveoli represent the essence of mass transfer in the body – small size and large overall surface area, the nephron is quite different from an alveolus. The alveolus is solely responsible for mass transfer of gases in one dimension and one direction. Mass transfer in the nephron includes the movement of ions, water, and metabolic waste products, such as urea, uric acid, and creatinine. Furthermore, the nephron initially removes more fluid and ions than the body can safely lose. It is the mechanism by which much of this fluid and ions are reabsorbed into the bloodstream that makes the nephron and the kidneys a unique mass transfer system. The nephron is seen in Figure 1.11. There are millions of nephrons in both kidneys.

The nephron consists of several segments. The initial filter is the glomerulus, which is embedded inside Bowman's capsule. This can be seen in Figure 1.12.

The glomerulus contains very large pores, each 50 angstroms in diameter and 500 angstroms long. These large pores allow a cumulative 125-150 ml/minute of filtered blood through to the remainder of the nephron. These pores allow water, ions, and metabolic waste products through, but they are too small to allow blood cells, proteins and large sugars to pass. Once the fluid passes through the glomerulus, it is then known as *filtrate*. Obviously, one cannot survive if 125 ml/minute of fluid were to leave the body. Thus, most of this fluid is reabsorbed back into the bloodstream, with the concentrated remainder resulting in urine. Normally, up to 124 ml are reabsorbed, resulting in a urine output of 1 ml/min, which is stored in the bladder.

The remainder of the nephron includes the proximal tubule, the loop of Henle, the distal tubule, and, finally, the collecting duct where urine is formed and collected. At each segment of the nephron, water and ions are removed from the tubules to be reabsorbed into the blood stream. The end result is more concentrated filtrate culminating in urine. Blood hormones control the amount of ions and water that are reabsorbed into the bloodstream, which results in more or less dilute urine. A more detailed description of mass transfer of ions and water from the tubules of the nephron is seen in Figure 1.13.

Parallel to the loop of Henle in the nephron is a loop of capillaries, called the *peritubular capillaries* or *vasa recta*. It is here where the water and ions in the filtrate are reabsorbed into the bloodstream. This parallel network is seen in Figure 1.14.

Mass transfer within the tubules of the nephron is a combination of concentration driven diffusion as well as active transport. The loop of Henle is configured in a counter-current approach, which allows for some substances that leave the descending loop of Henle to be reabsorbed into the ascending loop, in order to maximize the concentration of waste products, by controlling water transport. The water transport follows ion transport, so as to equalize the concentrations of both ions and water on either side of the tubule. Thus, the concentration at the bottom of the loop of Henle is very different from either top end (both descending – near the proximal tubule and ascending – near the distal tubule). This is shown in Figures 1.15 (a) and 1.15 (b).

The process by which mass transfer of waste products, water and excess ions are removed from a failing kidney is through the process of *dialysis*. Dialysis is a means of artificially cleaning the blood by accessing the radial artery for blood flow into a dialyzer cartridge embedded in a dialysis machine. The resulting clean blood is then returned through the cephalic vein into the body. There are thousands of cellulose capillary tubes within the dialysis cartriges, each with pores. Blood flows inside these capillary tubes, and a concentration driven diffusion occurs across the pores, which are

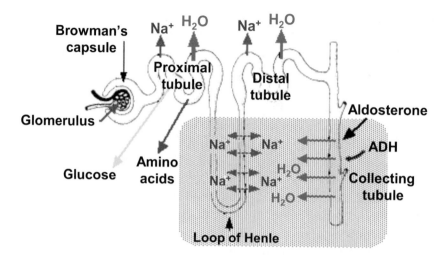

Figure 1.11: A nephron in the human kidneys (Wikipedia).

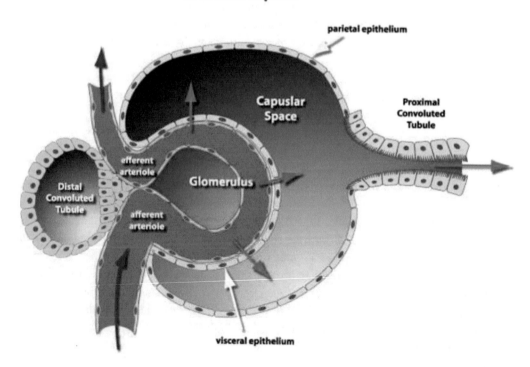

Figure 1.12: The glomerulus inside Bowman's capsule (Wikipedia).

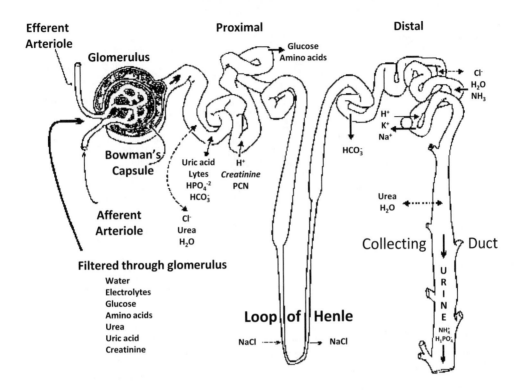

Figure 1.13: Mass transfer of ions, water and waste products along the length of the nephron and its segments.

large enough for transport of water, ions and the metabolic wastes, but they are too small for blood cells, proteins, or other vital blood components. Inside the cartridge is a fluid which accepts the transported substances. This fluid is called *dialysate*. The process of dialysis is shown in Figure 1.16.

The dialysis cartridge is shown in Figure 1.17, which depicts both the blood flow and the dialysate flow pathways.

The mass transport in the radial direction across the dialyzer cartridge capillary tube pores is a concentration driven diffusion process from blood to dialysate. Blood has a high concentration of wastes, and dialysate is waste free. Blood has a high concentration of ions, and dialysate has a normal concentration. This is seen in Figure 1.18.

The solubility of diffused substances within the capillary tube pore (labeled membrane in Figure 1.17) is different from that of the freely moving substances within blood and/or dialysate. Thus, there is a discontinuity in the concentration of substances within the membrane, which is

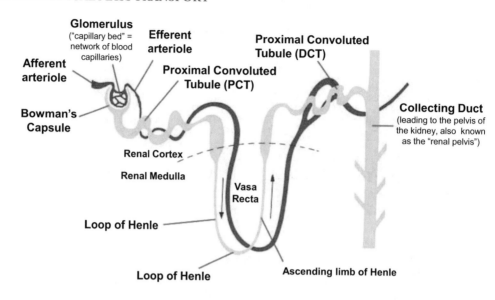

Figure 1.14: Vasa recta parallels the loop of Henle for reabsorption into the bloodstream.

noted by a membrane distribution coefficient. Note that the diffusion is from high concentration within the blood towards the lower concentration within the dialysate. There is also a free flow zone for both blood and dialysate (where the concentration is flat indicating no mass transfer) as well as a boundary layer near the wall of the capillary tube (where the concentrations are now altered due to mass transfer). The boundary layer produces slower axial flow rates, which allows for the radial mass transfer to proceed. The slower flow rates near the wall of the capillary tube within the boundary layer are due to shearing forces within flowing blood. This will be described in the subsequent section on fluid flow (momentum transport). The dialysate flow rate is typically much higher than the blood flow rate, in order to move the "dirty" dialysate (now having received the wastes and excess ions from the blood) out of the dialyzer cartridge and bring in clean dialysate. This keeps the concentration gradient for mass transfer large. A typical blood flow rate through the cartridge is 200 ml/min, while the dialysate flow rate is 500-800 ml/min.

Blood entering the top of the dialyzer cartridge is cleared of wastes and excess ions by the time blood leaves through the bottom of the cartridge. However, this blood is then returned to the bloodstream to be mixed with the remainder of the blood. Therefore, in order to completely clean the entire blood supply within the body, dialysis requires four hours to complete. Fick's law describes the concentration driven diffusion between blood and dialysate across the pores of the capillary tubes. As such, each layer of Figure 1.18 indicates a diffusivity (D), a thickness of the diffusion zone (∇X) and a concentration gradient ($C_b - C_b^*$), which are all factors associated with one-dimensional Fick's Law. Each layer has its own diffusivity, as D is a function of both the substance to be transported

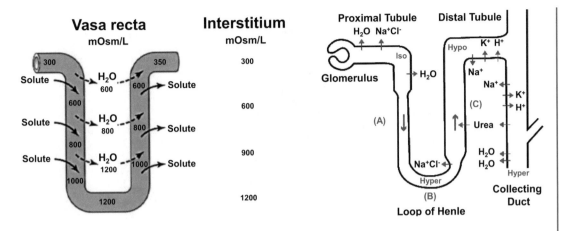

Figure 1.15: (a)-left, Changes in osmolarity. (b)-right, Counter-current system.

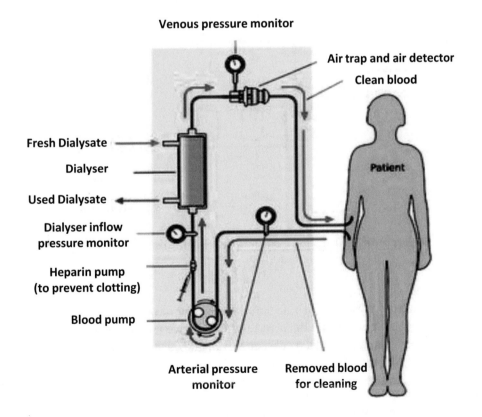

Figure 1.16: Dialysis system.

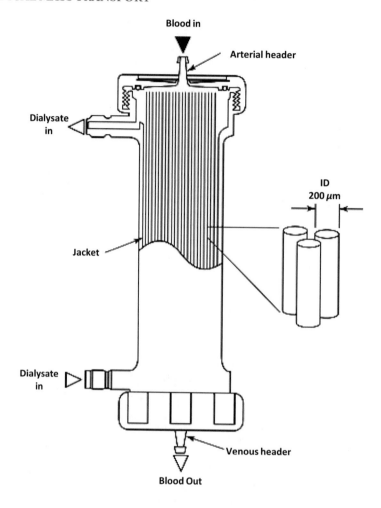

Figure 1.17: Dialyzer cartridge depicting blood and dialysate pathways.

along with the material through which it moves. Each layer has its own diffusion thickness and concentration gradient. It is true that the mass transfer of a given substance is the same for all three layers, as there is no accumulation of any substance in an individual layer. Therefore, an overall mass transfer rate across all three layers can be derived as is noted on the bottom of Figure 1.18, which uses the overall concentration gradient but incorporates the individual diffusivities and individual thicknesses.

Water transport via dialysis cannot be conducted by means of concentration driven diffusion, as the water concentration gradient is in the wrong direction (from dialysate towards blood). Therefore,

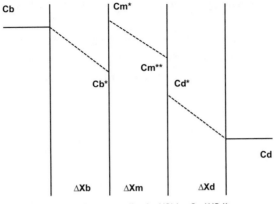

Cb
Cm*
Cm**
Cb*
Cd*
Cd
ΔXb
ΔXm
ΔXd
Φ = membrane distribution coeff = Cm*/Cb* = Cm**/Cd*
Mass flux across all layers = (Cb – Cd)/(ΔXb/Db + ΔXm/ΦDm + ΔXd/Dd)

Figure 1.18: Diffusion of wastes, ions and water in the radial direction across a dialysis capillary tube pore.

Water Removal Rate in Artificial Dialysis

$$Q = AL\,(\Delta P - \Delta p)$$

Q : water removed per hour (ml/hour)

A : membrane surface area (m^2)

L : ultrafiltration index (ml/[h-m^2-torr])(3 - 15 typical)

ΔP : mean hydrostatic pressure difference across membrane (140 torr, typical)

Δp : osmotic pressure difference across membrane (25 torr, typical)

Figure 1.19: Water removal during dialysis via a hydrostatic pressure gradient and an osmotic pressure gradient.

water transport from blood to dialysate is controlled by means of a hydrostatic pressure gradient, as noted in Figure 1.19.

The water transport is actually controlled via the dialysate flow rate, per the Bernoulli effect, similar to the manner by which a vacuum cleaner works. By having a large flow rate near an opening at right angles to that flow, there is a vacuum created at the opening, thus pulling the water out of the blood into the dialysate.

CHAPTER 2

Biofluid Mechanics and Momentum Transport

We have already seen examples of mass transport of substances across membranes within the human body and through those of artificial organs, such as dialyzers. In addition to mass transport, there are numerous examples of *momentum transport*, with momentum being the product of mass times velocity. This momentum transport is manifested in fluid, such as air flow in the lungs and blood flow through the human circulatory system. Much as there were guiding equations, such as Fick's law to govern mass transfer, there are also guiding equations to govern fluid flow. In addition, much as there were parameters which affected mass transfer and were properties of a given materials (diffusivity is an example), such is also the case for momentum transfer/fluid mechanics. Like diffusivity is a proportionality constant for mass transfer, so is *viscosity* a proportionality constant for fluid mechanics. In mass transfer, the driving force was either a concentration gradient (for liquids or solutes embedded in solvents) or a partial pressure gradient (for gases). For fluid flow and momentum transport, the driving force is a pressure gradient. This is not a partial pressure gradient but rather a standard pressure gradient, often called the hydrostatic pressure.

However, the total pressure gradient for a flowing fluid is the sum of the hydrostatic pressure, the dynamic pressure (related to the fluid velocity) and the pressure associated with gravity and a height differential. The gravity component is easily described by the height difference in a fluid manometer. An example of a manometer, which can be used to measure a pressure difference as a result of the density and gravitational difference between two fluids, is shown in Figure 2.1 below.

The total pressure of a flowing fluid is given by:

$$P_T = P \text{ (hydrostatic)} + 1/2\, \rho V^2 + \rho g h .$$

Where the hydrostatic pressure is what is normally though of as "pressure," the dynamic pressure is given by $1/2\, \rho V^2$ and the last term is the pressure as a result of a height differential and is measured by a manometer. A pressure transducer can be used to measure the hydrostatic pressure, a flowmeter is used to measure the dynamic pressure, and a manometer for the gravity term. Depending on the density of the fluid and the value for fluid velocity, it is possible that one of the three pressure components can far outweigh the others. The Bernoulli equation is a modification of the total pressure equation. It states that the total pressure at two points is the same, or that the sum of the three pressures is equal.

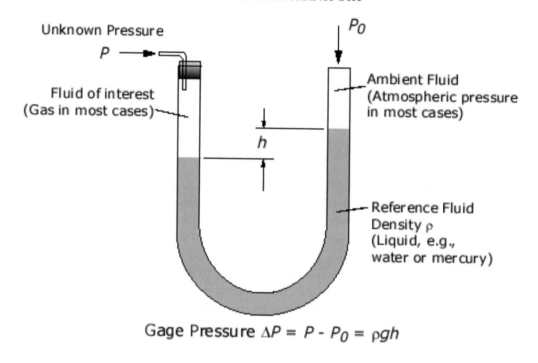

Gage Pressure $\Delta P = P - P_0 = \rho g h$

Figure 2.1: Fluid manometer used to measure the density/gravitational pressure difference between two fluids with varying heights within the manometer. The pressure gradient is a function of the density multiplied by the height difference in the two fluids.

2.1   BLOOD AND FLUID VISCOSITY

Viscosity is a measure of the *resistance* of a flowing *fluid* that is being deformed by either *shear stress* or *normal stress*. In general, it is the resistance of a liquid to flow, or its "thickness." Viscosity describes a fluid's internal resistance to flow and may be thought of as a measure of fluid *friction*. Thus, *water* is "thin," having a lower viscosity, while *vegetable oil* is "thick," having a higher viscosity. All real fluids (except *superfluids* such as liquid nitrogen) have some resistance to *stress* and fluid flow, but a fluid which has no resistance to shear stress is known as an *ideal fluid* or *inviscid* fluid. A real fluid with a non-zero viscosity is called a *viscous* fluid. The study of viscosity of a real fluid is known as *rheology*.

Isaac Newton postulated that, for straight, *parallel* and uniform flow, the shear stress, τ, between fluid layers is proportional to the *velocity gradient*, $\partial u/\partial y$, in the direction *perpendicular* to the layers.

$$\tau = \mu \frac{\partial u}{\partial y} \; .$$

Here, the constant μ is known as the *coefficient of viscosity*, the *viscosity*, the *dynamic viscosity*, or the *Newtonian viscosity*. Many *fluids*, such as *water* and most *gases*, satisfy Newton's criterion and are known as *Newtonian fluids*. Non-Newtonian fluids exhibit a more complicated relationship between shear stress and velocity gradient than simple linearity. Many such Non-Newtonian fluids have a high concentration of suspended particles. Blood may be one such fluid, as blood contains almost 50% of its volume as blood cells. In the metric system, also known as the cgs (centimeter-gram-second) system, the unit of viscosity is the *Poise*, named after *Jean Louis Marie Poiseuille* who formulated *Poiseuille's law* of viscous flow. The value of viscosity is so small for many fluids (liquids and gases) that their viscosity is often listed in cP, centipoises, or one hundredth of a poise. A poise is equal to one dyne/cm-sec.

As an example, the viscosity of water is approximately 1 cP with a density of 1 g/cc. Blood has a viscosity of 3-4 cp, with a density of 1.05 g/cc. As can be seen, although the densities of water and blood are fairly similar, their viscosities are not. This is due to the large suspension of cells within blood, which adds to the resistance to blood flow. The more solid/liquid boundaries in a fluid, where the shearing stresses tend to be higher, the higher the viscosity. The cells within blood add numerous "solid"/liquid boundaries. Denser pure fluids, such as honey and molasses are thicker fluids with a much higher viscosity than water or blood. When a pure liquid (without a suspension of particles) has a higher density, its viscosity also tends to be proportionally higher.

Fluids with large suspensions, such as blood, have viscosities that are proportional to the concentrations of particles. For blood, this is the hematocrit, which is the percentage of red cells by volume in whole blood. Normally, the hematocrit is 45%. Figure 2.2 depicts the relationship between blood hematocrit and viscosity.

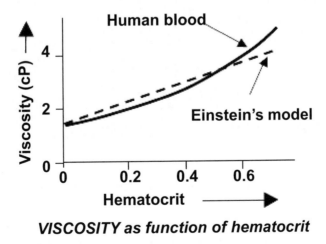

Figure 2.2: Blood viscosity as a function of blood hematocrit. Albert Einstein developed a model of viscosity of a fluid with a large concentration of substances where $c > 10\%$.

The blood consists of a suspension of cells in a liquid called *plasma*. In an adult man, the blood is about 1/12th of the body weight and this corresponds to 5-6 liters. Blood consists of 55% plasma, and 45% by cells, sometimes called *formed elements*. The blood performs many important functions. By means of the hemoglobin contained in the erythrocytes (red cells), it carries oxygen to the tissues and collects the carbon dioxide. It also conveys nutritive substances (e.g., amino acids, sugars, mineral salts) and gathers the excreted material, which will be eliminated through the kidney. The blood also carries hormones, enzymes and vitamins. It performs the defense of the body by mean of the phagocitic activity of the leukocytes (white cells).

The erythrocytes are the most numerous blood cells: about 4-6 millions/mm^3. They are also called red cells. In man and in all mammals, erythrocytes are devoid of a nucleus and have the shape of a biconcave lens. The red cells are rich in hemoglobin, a protein able to bind to oxygen. Hence, these cells are responsible for providing oxygen to tissues and partly for recovering carbon dioxide produced as waste. However, most CO_2 is carried by plasma, in the form of soluble carbonates. Red cells, which are the most prolific cells in the bloodstream, form the basis of the hematocrit, which is a factor in the determination of blood viscosity. A typical red cell is seen in Figure 2.3.

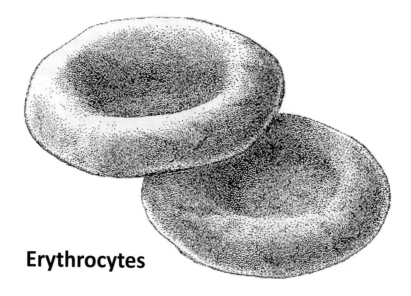

Erythrocytes

Figure 2.3: Red cells consisting of biconcave disks, each 8 microns in diameter and 2 microns thick.

The measurement of viscosity is conducted with a device known as a *viscometer*. For liquids with viscosities, which vary with flow conditions, an instrument called a *rheometer* is used. Viscometers only measure viscosity under one flow condition. In general, either the fluid remains stationary and an object moves through it, or the object is stationary and the fluid moves past it. The drag caused by relative motion of the fluid and a surface is a measure of the viscosity.

A falling sphere viscometer is one of the simpler types of viscometers. It is based on Stoke's Law, which describes flow of a fluid past a sphere in a vertical tube, where the flow is very slow. A sphere of known size and density is allowed to descend through the liquid. If correctly selected, it reaches *terminal velocity*, which can be measured by the time it takes to pass two marks on the tube. Knowing the terminal velocity, the size and density of the sphere, and the *density* of the liquid, Stokes' law can be used to calculate the *viscosity* of the fluid. A series of steel ball bearings of different diameter is normally used in the classic experiment to improve the accuracy of the calculation. Stoke's Law is given as:

$$F = 6\pi r \eta v$$

where:
F is the frictional force, r is the radius of the spherical object, η is the fluid viscosity, and v is the particle's velocity. An example of a functional relationship for a capillary tube viscometer is shown in Figure 2.4.

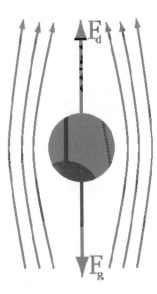

Figure 2.4: Slow flow past a falling sphere, which forms the basis of a falling sphere viscometer.

Since blood is opaque, sensors within the tube indicate when the sphere passes points along the tube. Another type of viscometer is the *Couette* or concentric cylinder viscometer. The fluid is placed between two cylinders with large radii and with a gap width between the cylinders, which is very small compared to the radii. One of the cylinders rotates while the other is stationary. With a small gap width, the velocity between the cylinders approaches a linear profile. The torque required to spin the cylinder as a function of the cylinder rotation rate provides the fluid viscosity. Different fluids would have different linear flow profiles between the cylinders, indicating different values of viscosities. A typical concentric cylinder viscometer is seen in Figure 2.5.

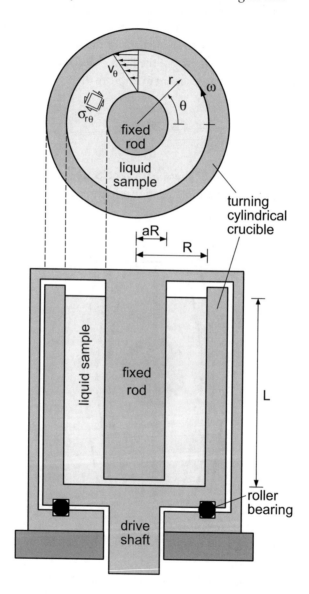

Figure 2.5: Concentric cylinder viscometer.

2.2   CONSERVATION OF MASS

The equation which describes fluid flow as a function of the pressure gradient in a tube is used to describe blood flow in a blood vessel or air flow in the trachea and bronchioles. As was the case with diffusion occurring in the direction of decreasing concentration gradient, so is the case of fluid flow in the direction of decreasing pressure in a blood vessel. In the human systemic circulation, the aortic pressure is 120 systolic pressure over 80 diastolic pressure (in mm Hg). This is because blood flow is pulsatile coming from the aortic valve as the ventricles undergo systolic contraction, followed by diastolic relaxation. The blood pressure is 100 mm Hg mean and eventually decreases to @ 5-10 mm Hg at the vena cava. Thus, there is a 90-95 mm Hg pressure gradient, which pushes blood around the systemic circulation. Blood flow in the aorta is highly accelerated and time varying and continues around the sharp bend of the aortic arch. The equations that describe blood flow and fluid mechanics are time varying in nature, as well as affected by spatial variation due to the changing geometry/anatomy of the circulatory system. These equations include the equation relating to the conservation of mass (sometimes called the continuity equation) and the equations of fluid momentum (sometimes called the equations of fluid motion).

As was noted above, the type of fluid (Newtonian or non-Newtonian) can affect the equations related to fluid mass and fluid momentum. For simplicity, we can assume that blood acts Newtonian and can then alter that assumption after deriving the equations shown above. The continuity equation for a Newtonian fluid is shown below. We are using the version of this equation in cylindrical coordinates as blood vessels are cylinders. Thus, the velocities associated with this coordinate system are in the z direction (axially downstream), the r direction (radially from the center to the edge of the vessel) and in the theta direction (circumferentially). This equation assumes that the density is a constant. A constant density assumes the fluid to be *incompressible*, which is a realistic assumption for most liquids (such as blood) and is unrealistic for most gases, which can compress at higher pressures.

$$\frac{1}{r}\frac{\partial}{\partial r}(ru_r) + \frac{1}{r}\frac{\partial u_\theta}{\partial \theta} + \frac{\partial u_z}{\partial z} = 0.$$

2.3   FLUID MOMENTUM, THE EQUATIONS OF MOTION, AND THE NAVIER–STOKES EQUATIONS

The Navier–Stokes equations, named after *Claude–Louis Navier* and *George Gabriel Stokes*, describe the motion of *viscous fluids* such as *liquids* and *gases*, and therefore the conservation of fluid momentum. These equations arise from applying *Newton's second law* to *fluid* motion, together with the assumption that the fluid *stress* (τ in the section above describing viscosity) is the sum of a *diffusing viscous* term (proportional to the gradient of velocity), plus a *pressure* term. As was the case for the continuity equation, the Navier–Stokes equations may be written in various coordinates systems, but the cylindrical coordinate system is most appropriate for blood flow, as blood vessels are cylinders.

$$\rho\left(\frac{\partial u_r}{\partial t} + u_r\frac{\partial u_r}{\partial r} + \frac{u_\theta}{r}\frac{\partial u_r}{\partial \theta} + u_z\frac{\partial u_r}{\partial z} - \frac{u_\theta^2}{r}\right)$$

$$= -\frac{\partial p}{\partial r} + \mu\left[\frac{1}{r}\frac{\partial}{\partial r}\left(r\frac{\partial u_r}{\partial r}\right) + \frac{1}{r^2}\frac{\partial^2 u_r}{\partial\theta^2} + \frac{\partial^2 u_r}{\partial z^2} - \frac{u_r}{r^2} - \frac{2}{r^2}\frac{\partial u_\theta}{\partial\theta}\right] + \rho g_r$$

$$\rho\left(\frac{\partial u_\theta}{\partial t} + u_r\frac{\partial u_\theta}{\partial r} + \frac{u_\theta}{r}\frac{\partial u_\theta}{\partial \theta} + u_z\frac{\partial u_\theta}{\partial z} + \frac{u_r u_\theta}{r}\right)$$

$$= -\frac{1}{r}\frac{\partial p}{\partial\theta} + \mu\left[\frac{1}{r}\frac{\partial}{\partial r}\left(r\frac{\partial u_\theta}{\partial r}\right) + \frac{1}{r^2}\frac{\partial^2 u_\theta}{\partial\theta^2} + \frac{\partial^2 u_\theta}{\partial z^2} + \frac{2}{r^2}\frac{\partial u_r}{\partial\theta} - \frac{u_\theta}{r^2}\right] + \rho g_\theta$$

$$\rho\left(\frac{\partial u_z}{\partial t} + u_r\frac{\partial u_z}{\partial r} + \frac{u_\theta}{r}\frac{\partial u_z}{\partial \theta} + u_z\frac{\partial u_z}{\partial z}\right)$$

$$= -\frac{\partial p}{\partial z} + \mu\left[\frac{1}{r}\frac{\partial}{\partial r}\left(r\frac{\partial u_z}{\partial r}\right) + \frac{1}{r^2}\frac{\partial^2 u_z}{\partial\theta^2} + \frac{\partial^2 u_z}{\partial z^2}\right] + \rho g_z \ .$$

2.4   POISEUILLE FLOW

As can be seen above, even with the assumption of a Newtonian fluid, these equations are time varying with partial derivatives and are difficult to solve. As such, the Poiseuille assumptions, named after *Jean Louis Marie Poiseuille* who formulated *Poiseuille's law* of viscous flow (as mentioned above) are as follows:

1. The flow is steady ($\partial(\ldots)/\partial t = 0$).

2. The radial and swirl components of the fluid velocity are zero ($u_r = u_\theta = 0$).

3. The flow is axisymmetric ($\partial(\ldots)/\partial\theta = 0$) and fully developed ($\partial u_z/\partial z = 0$).

As a result, the first two of the three Navier–Stokes *momentum* equations and the *continuity equation* are identically satisfied. The third (z) momentum equation reduces to

$$\frac{1}{r}\frac{\partial}{\partial r}\left(r\frac{\partial u_z}{\partial r}\right) = \frac{1}{\mu}\frac{\partial p}{\partial z} \ .$$

There are two boundary conditions for the flow of a viscous fluid (such as blood):
$(dV_z/dr) = 0$ when $r = 0$ (center of blood vessel) and $V_z = 0$ at $r = R$ (edge of tube), which are the result of axisymmetric flow and viscous "no slip at the wall," respectively.

The solution is

$$u_z = \frac{1}{4\mu}\frac{\partial p}{\partial z}r^2 + c_1\ln r + c_2 \ .$$

Since u_z needs to be finite at $r = 0$, $c_1 = 0$. The no slip *boundary condition* at the pipe wall requires that $u_z = 0$ at $r = R$ (radius of the pipe), which yields

$$c_2 = -\frac{1}{4\mu}\frac{\partial p}{\partial z}R^2 \ .$$

Thus, we have finally the following *parabolic velocity* profile:

$$u_z = -\frac{1}{4\mu}\frac{\partial p}{\partial z}\left(R^2 - r^2\right).$$

The maximum velocity occurs at the pipe centerline ($r = 0$):

$$u_{z\,\text{max}} = \frac{R^2}{4\mu}\left(-\frac{\partial p}{\partial z}\right).$$

The blood flow is thus a paraboloid with a maximum velocity at the center of the blood vessel with a zero velocity at the wall, as seen in Figure 2.6. The velocity, varying in the radial direction, can be thought of as flowing within concentric layers, each with a different velocity at a different radial value. The zero value at the wall corresponds to the "no slip" condition associated with Poiseuille flow. The fluid velocity is symmetric about the center of the tube, corresponding to the axisymmetric assumption associated with Poiseuille flow.

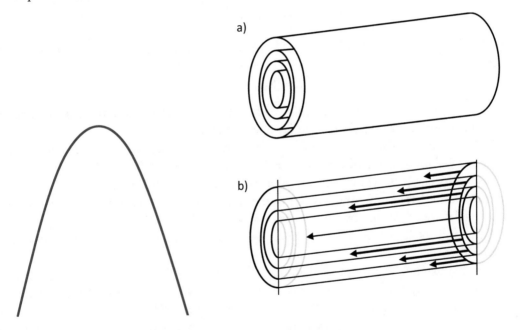

Figure 2.6: The parabolic velocity profile associated with the solution of the Navier–Stokes equations for Poiseuille flow. The velocity is maximal at the center of the vessel and zero at the wall. The velocity is axisymmetric corresponding to the associated Poiseuille assumption (Wikipedia).

If one integrates the velocity over the cross sectional area of the blood vessel, this results in the well known Poiseuille law for steady, incompressible, axisymmetric flow in a cylindrical vessel:

$$Q = pR^4 \nabla P/8\eta l$$

where Q is the volume flow rate (cardiac output for blood), R is the vessel radius, ∇P is the pressure gradient, η is the blood viscosity, and l is the length of the blood vessel over which the pressure gradient is acting.

Poiseuille's law (sometimes called the Hagen-Poiseuille law) states that, assuming *laminar flow* in a tube, flow (Q) is proportional to its radius to the 4th power (R^4) and inversely proportional to its length (l) and viscosity of the fluid (η). The general principle of this law is that small changes in the internal diameter of a blood vessel *lumen* can make a big difference to the rate of flow. The flow in the aorta is large, in part because the radius of the aorta is large. The flow in smaller vessels is less, not only because the branching blood vessels segment the available flow, but also because the vessel radius is smaller. As stated above in the fluid viscosity section, when the hematocrit is smaller, so is the blood viscosity. As a result, an anemic patient has a smaller blood viscosity; therefore, according to the Poiseuille Law, the flow rate is larger. The resistance to blood flow (Poiseuille) corresponds to the relationship between volume flow rate (Q) and the pressure gradient (∇P) by the equation

$$Q = \nabla P / R$$

where R is the resistance to flow. If one uses the Poiseuille flow equation shown above, then the flow resistance takes the form of

$$R = 8\eta l / p R^4 .$$

Thus, the resistance to flow inside a given blood vessel is proportional to the fluid viscosity and inversely related to the vessel radius to the fourth power.

Throughout the analysis of fluid flow in a vessel, the driving mechanism has been the pressure gradient (∇P). An example of the relationship between pressure and flow is the relationship between left ventricular pressure in the heart, left atrial pressure, and aortic pressure. The heart valves associated with the left heart (mitral valve between the left atrium and left ventricle, and aortic valve between the left ventricle and the aorta) open and close via the difference in pressure across each valve. This is evident in Figure 2.7, which shows the pressure and timing relationship across the two left heart valves.

Another manifestation of pressure drop producing a fluid flow is the relationship between pressure across the systemic circulation, resulting from flow across the aortic valve, through the aorta, arteries, arterioles, capillaries, venules, veins, and the vena cava. The pressure across the systemic circulation begins as a pulsatile pressure in the aorta (at a mean pressure @ 100 mm Hg) and ends at the vena cava with a mean pressure of @ 5 mm Hg. This is shown in Figure 2.8 along with the smaller pressures associated with the right heart and the pulmonary circulation.

2.5 BOUNDARY LAYERS

The Poiseuille assumption of fully developed flow ($\partial u_z/\partial z = 0$) refers to the fluid boundary layer, the region of the flow field where viscous forces predominate. What is "fully developed" is the boundary layer. Thus, for Poiseuille flow, the entire flow field is viscous dominated. This is manifested by the

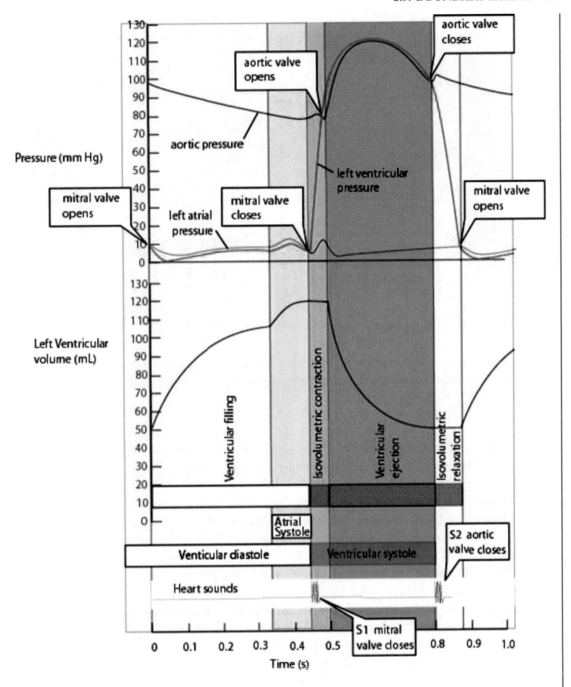

Figure 2.7: Pressure relationships across the left heart which produce flow through the left atrium and left ventricle (across the mitral valve) or though the left ventricle and aorta (across the aortic valve).

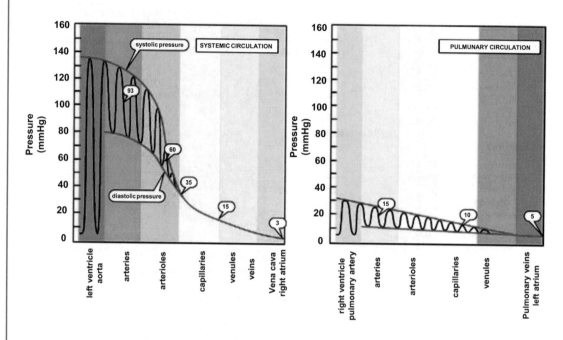

Figure 2.8: Pressure reduction across the systemic circulation (left side) and the pulmonary circulation (right side) indicating the continual pressure gradient which drives blood flow along the circulatory systems.

parabolic flow profile mentioned above. However, in reality, the fluid flow in the aorta is not truly parabolic, but is more of a blunt profile. The flow profile becomes more rounded as the boundary layer grows from the outside of the vessel towards the center.

From a momentum transport perspective, the outside of the blood vessel, where the boundary layer is growing, is said to undergo convective deceleration, while in the center of the vessel, the flow is undergoing convective acceleration. Thus, there is a transfer of fluid momentum from the outside of the vessel towards the inside, until, finally, the entire flow pattern is completely rounded and parabolic. The growth of the boundary layer is shown in Figure 2.9 for *laminar flow*, which is flow of a fluid that is smooth and orderly.

The boundary layer thickness is given by the symbol d whose value is quite small during "plug" flow at the entrance to a blood vessel, but which slowly grows during viscous flow until it equals the vessel radius. At that point, the velocity profile is fully parabolic, and the flow is said to be "fully developed." This is the point where the Poiseuille assumptions are now valid. *Please note that the pulsatile nature of blood flow in an artery has been neglected in this analysis and in Figure 2.9.*

When the fluid flow is *turbulent*, the flow profile is more chaotic. This results in more fluid mixing from the areas of higher velocity flow at the center with lower velocity flow at the periphery

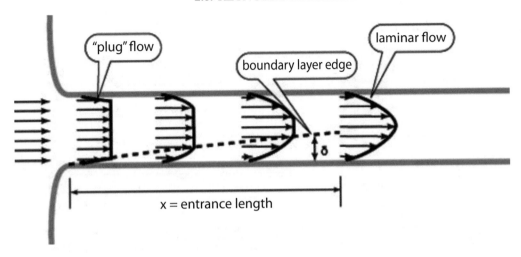

Figure 2.9: Growth of the laminar boundary layer from a blunt profile on the left towards a more rounded profile at the right, where the boundary layer becomes fully developed.

of the vessel (towards the blood vessel wall). Thus, turbulent flow has more mixing and therefore more momentum transport between the center of the vessel and the wall. In laminar flow, the fluid shear stress is highest at the wall and zero at the center. Even so, the laminar wall shear stress is not huge. However, in turbulent flow, the higher velocities are convected towards the wall and the wall shear stress is higher than that of laminar flow. The turbulent boundary layer grows at a different rate than for laminar flow and the fully developed flow profile is not parabolic, but more blunt in shape. This is seen in Figure 2.10.

If the boundary layer separates from the wall, there is a zone of stagnant fluid. For blood, this would be a zone where *thrombosis* (clotting) would occur. Such a clot might become dislodged where it would be transported downstream to lodge in a smaller blood vessel. This might result in a stroke, if the dislodged clot (called an *embolism*) were to become trapped in the cerebral (brain) circulation. Such boundary layer separation might occur at sharp bends where the fluid flow cannot adequately follow the curvature. This is seen in Figure 2.11.

Boundary layer separation might also occur at branches of blood vessels as is seen in Figure 2.12.

2.6   REYNOLDS NUMBER AND TYPES OF FLUID FLOW

The differences between laminar and turbulent flow are considerable. Laminar flow, sometimes known as *streamline* flow, occurs when a fluid flows in parallel layers, with no disruption between the layers. It is the opposite of *turbulent flow*. In nonscientific terms, laminar flow is "smooth" while turbulent flow is "rough." The *dimensionless Reynolds number* is an important parameter in

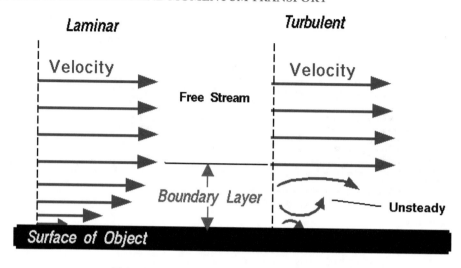

Figure 2.10: Laminar and turbulent boundary layers with laminar version more rounded than turbulent version. Turbulent version has greater mixing and greater momentum transport.

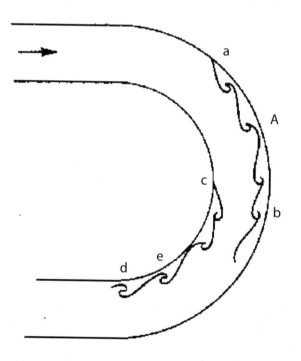

Figure 2.11: Boundary layer separation at a tight curve, such as the aortic arch.

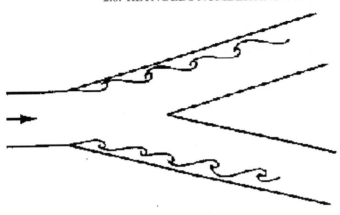

Figure 2.12: Boundary layer separation in branching blood vessels.

the equations that describe whether flow conditions lead to laminar or turbulent flow. It indicates the relative significance of the viscous effect compared to the inertia effect. *The Reynolds number* is proportional to the ratio of the inertial force (acceleration) to the viscous force (fluid deceleration). The values of the Reynolds number for various types of flow are shown below:

- **laminar** if $Re < 2000$

- **transient** if $2000 < Re < 3000$

- **turbulent** if $3000 < Re$

These are approximate values. The Reynolds number can be affected by the anatomy/geometry of the fluid flow field, the roughness of the vessel wall, and irregularities in pressure or external forces acting on the fluid. When the Reynolds number is much less than 1, *creeping motion* or *Stokes flow* occurs. This is an extreme case of laminar flow where viscous (friction) effects are much greater than the virtually non-existent inertial forces. Stokes flow was described above regarding the falling sphere viscometer.

As was noted above, the Reynolds number is dimensionless and gives a measure of the *ratio* of *inertial forces* ($V\rho$) to *viscous* forces (μ/L); consequently, it quantifies the relative importance of these two types of forces for given flow conditions. The equation for the Reynolds number is:

$$Re = \frac{\rho V D}{\mu} = \frac{V D}{\nu} = \frac{Q D}{\nu A} \; .$$

where

- V is the mean fluid velocity in (cm/s).

- D is the diameter (cm).

- μ is the *dynamic viscosity* of the *fluid* (g/cm-sec).

- ν is the *kinematic viscosity* ($\nu = \mu/\rho$) (cm^2/s).

- ρ is the *density* of the fluid (g/cm^3).

- Q is the *flow rate* (cm^3/s).

- A is the pipe *cross-sectional* area (cm^2).

The first form of the equation is the most common form. In the circulatory system, the Reynolds number is 3000 (mean value) and 7500 (peak value) for the aorta, 500 for a typical artery, 0.001 in a capillary, and 400 for a typical vein. This corresponds to turbulent flow in the aorta (where the aortic wall is strengthened to overcome the turbulent forces), laminar flow in arteries and veins, and creeping flow in capillaries. Thus, there are added forces, mixing and momentum transport in the aorta and virtually no momentum transport in capillaries. However, as was noted in the discussion on mass transfer in systemic capillaries, it is the low axial blood flow that allows for the radial mass transfer to occur. This could not happen in arteries, as the fluid velocity is too large to allow any significant radial mass transfer to occur. However, with turbulent flow in the aorta, there is considerable mixing of fluid layers and added wall shear stresses. As a result, there is added mass transfer across the artery walls, which is why arterial disease is prominent in the aorta. Such a disease is caused by mass transfer of lipids across the arterial wall and by distortion of the endothelial cells lining the vessel wall, allowing for such mass transfer to occur in the gaps.

A depiction of the systemic capillaries is shown in Figure 2.13. The Reynolds number is less than 1 and the flow is creeping flow. Such flow produces entirely viscous flow with no inertial effects, no turbulence, and no flow separation. As such, there are no negative effects due to the circuitous nature of the capillary bed nor to sudden sharp turns or bifurcations. In larger vessels, such vessel geometry would produce turbulence and/or flow separation.

2.7 BLOOD PRESSURE MEASUREMENT

The measurement of blood pressure non-invasively is conducted by the well known blood pressure cuff assembly, the sphygmomanometer. Most individuals often have had their blood pressure taken via this mechanism. It involves inflating a pressure cuff just above the elbow to a pressure significantly exceeding systolic blood pressure (120 mm Hg). At such a high pressure, the artery just below the surface (brachial artery) collapses as the externally applied pressure from the cuff exceeds the internal blood pressure inside the artery. A stethoscope is paced just below the cuff. As the pressure

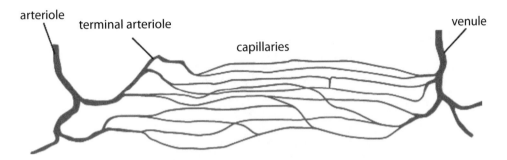

Figure 2.13: The network of the systemic capillaries leading from arterioles and ending in venules, with bending and bifurcating elements in the capillary bed that are possible with creeping flow.

is reduced, the external pressure will eventually fall slightly below the systolic blood pressure, and the brachial artery will open slightly. The blood at that high pressure will jet through the small opening producing turbulent flow. Such jetting turbulent flow can be heard via the stethoscope. This is similar to a heart murmur heard through a stethoscope, which is indicative of a leaky heart valve, also producing turbulence. As the cuff pressure continues to fall, the duration that the brachial artery pressure exceeds the cuff pressure will increase. As such, the (small) opening will increase in size and duration, until finally the vessel remains open as the cuff pressure falls below diastolic pressure in the artery. At that point, the turbulent sounds, called Korotkoff sounds, fade away.

A manometer is connected to the pressure cuff so that the physician can see what the pressures are when the sound begins (systolic pressure) and when the sound finally fades away (diatolic pressure). Thus, the sphygmomanometer can be used as a non-invasive means of determining the two extremes of blood pressure, normally at 120/80 mm Hg. The relationship between cuff pressure, vessel pressure, and sounds via the stethoscope are shown in Figure 2.14.

The measurement of blood pressure depends greatly on the location of the measurement. As was seen in Figure 2.8, the blood pressure falls as blood circulates through the systemic circulatory system. However, even though blood pressure is normally taken in the arterial network, the location of that measurement can affect the systolic and diastolic values. As can be seen in Figure 2.15, the blood pressure waveform can be steeper with a corresponding larger difference between systolic and diastolic pressure, depending upon the location within one of the mid-sized arteries. The increase in the pulse pressure (difference between systolic and diastolic) is sometimes due to a reflection of the pressure *wave*, which is the propagation of the pulse wave (moving far faster than the fluid time varying pressure).

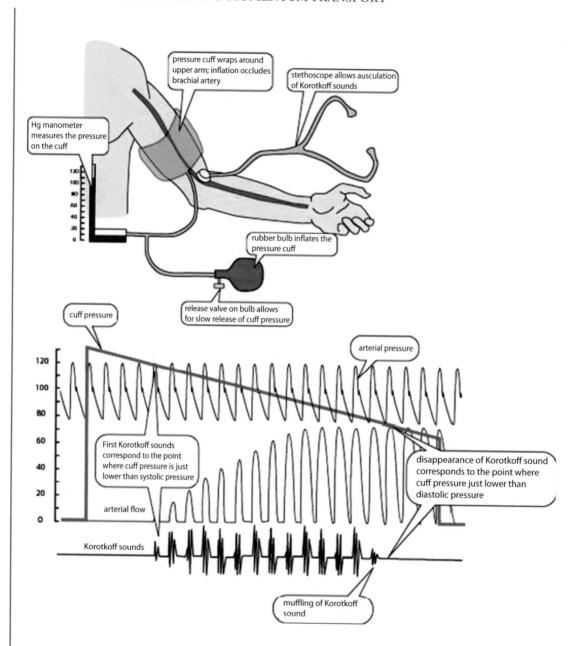

Figure 2.14: The sphygmomanometer and the generation of Korotkoff sounds indicating systolic and diastolic arterial pressure.

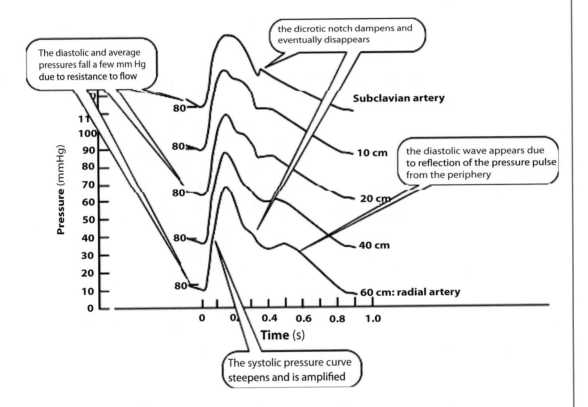

Figure 2.15: The blood pressure waveform and the resulting pulse pressure (difference between systolic and diastolic) can vary in differing mid-level arteries. This would result in a different measured systolic and diastolic values of blood pressure via a sphygmomanometer.

CHAPTER 3

Biomedical Heat Transport

Heat transport within the human body provides the mechanism for the body's core temperature to remain constant, by shunting heat from the core to the periphery, where heat is lost through the skin and through respiration. Heat is generated inside the body as it metabolizes food, which produces heat as a by-product. Heat transfer from the body to the environment is also affected by the outside temperature and wind speed. The latter affects wind chill, which can produce a substantial heat loss from the body. Heat transport also occurs in blood heating and cooling devices, which are often used during surgical procedures and as part of heart-lung machines. Finally, heat transport is vital for extreme conditions such as a fireman's suit or a spacesuit, both used to protect the body from extreme temperatures in the environment.

The heat generated by the body at rest is called the *basal metabolic rate*, equal to 72 kcal/hr. Obviously, as metabolic activity increases, such as during exercise, the level of heat production rises.

The equations describing heat transport are similar in form to the equations relating transport of mass (Fick's Law) and momentum. Such transport is always in the direction of a decreasing gradient. For mass, the gradient was the concentration difference between two points or across membranes. For momentum transport of fluids, it was the pressure differential causing fluid flow. For heat transfer, the driving gradient is the temperature difference between two points.

3.1 CONDUCTION, CONVECTION, AND RADIATION – THE TYPES OF HEAT TRANSFER

Heat transport can occur by three basic mechanisms. Thermal *conduction* is the process by which heat transfer occurs by molecular interaction. It can occur in gases, liquids and solids. An example is heat transfer across a closed window or across a wall. In the human body, the example of conductive heat transfer is across tissue from the core towards the periphery (skin). Thermal *convection* is the process by which heat transfer occurs via bulk motion of a fluid. An example might be forced air flow from an air conditioner vent or wind chill on a windy day. In the body, examples of convective heat transfer include blood flow from the core towards the periphery or air flow in the lungs from the alveoli though the trachea and out the mouth and nose. Both conduction and convection require that there be a material involved, although convection does not occur in solids, whereas conduction can. The third method of heat transfer is thermal *radiation*, which occurs as a result of electromagnetic transport processes. An example is heat gain from the sun to the earth (and your own skin). Radiation is a surface to surface phenomenon and does not require a material interaction. The greenhouse effect is an example of thermal radiation.

As an example, Figure 3.1 depicts the three types of heat transfer from a campfire.

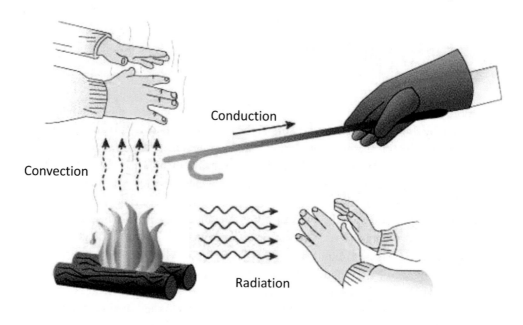

Figure 3.1: The three types of heat transfer: conduction, convection and radiation (Wikipedia).

3.2   THERMAL CONDUCTION

Conduction (or heat conduction) is the transfer of *thermal energy* between neighboring molecules in a substance due to a *temperature gradient*. It always takes place from a region of higher temperature to a region of lower temperature, and it acts to equalize temperature differences. Conduction takes place in all forms of matter including solids, liquids, and gases, but it does not require any bulk motion of matter. In solids, it is due to the combination of vibrations of the molecules in a *lattice* and the energy transport by *free electrons*. In gases and liquids, conduction is due to the collisions and *diffusion* of the molecules during their random motion. Conduction is the movement of heat from a warmer object to a cooler one when they are in direct contact with one another. This serves to even the temperature difference between them over time. The rate of heat transfer between two objects of different temperatures depends upon several factors. These include

- The temperature difference between the two objects.

- The total surface area where the two objects are in contact.

- The efficiency of the insulation that is between the objects.

The greater the temperature difference between two objects in contact, the more heat is transferred between them in a given time. For example, when you place your hand on a very hot stove top, you will quickly receive a great heat input from the stove to your hand. If the stove top is only warm, it will take much longer to receive the same amount of heat into your hand.

The more surface area in contact between two objects, the more quickly heat is transferred between them. Stick your finger on an icicle for a minute, and it feels cold, but you will probably not feel too uncomfortable. Strip naked and lay on a block of ice for a minute, and you will most likely be very uncomfortable, indeed, as the ice absorbs heat from your body at a very fast rate.

The amount of heat being transferred between two objects of different temperatures can be slowed by the use of effective insulation. Insulation retards the movement of heat between them by creating pockets of dead air space that trap the flow of heat or by otherwise slowing the overall heat transfer rate by adding a low conductance/high resistance layer.

The law of Heat Conduction, also known as *Fourier's* law, states that the time rate of *heat transfer* through a material is *proportional* to the negative *gradient* in the temperature and to the area at right angles, to that gradient, through that the heat is flowing. We can state this law in two equivalent forms: the integral form, in which we look at the amount of energy flowing into or out of a body as a whole, and the differential form, in which we look at the flows or *fluxes* of energy locally. The differential form is the following:

$$\vec{q} = -k\nabla T$$

where

$\vec{q}$ is the local *heat flux*, $[W \cdot m^{-2}]$,

κ is the material's *conductivity*, $[W \cdot m^{-1} \cdot K^{-1}]$, and

∇T is the temperature gradient, $[K \cdot m^{-1}]$.

The term "K" refers to degrees Kelvin, an absolute scale for temperature. For materials with a high thermal conductivity, such as metals, there is little if any insulating capacity. For materials with a low thermal conductivity, the material has more insulating characteristics. Air and many gases have low thermal conductivities. Human tissue also has a relatively low thermal conductivity and is thus relatively insulating, which aids in reducing heat transfer to the core of your body on extremely hot or cold days. Clothing helps to further insulate the body by providing extra layers for heat conduction and by trapping air, which is itself a thermal insulator.

For many simple applications, a one-dimensional form of Fourier's law is written as:

$$q_x = -k\frac{dT}{dx} \ .$$

The thermal conductivity, k, is often treated as a constant, though this is not always true. While the thermal conductivity of a material generally varies with temperature, the variation can be small over a significant range of temperatures for some common materials. In anisotropic materials, the thermal conductivity typically varies with orientation. For the human body, trapped air in the lungs has a different conductivity as opposed to human tissue, but both have smaller values than more dense materials, such as human bone. Water (and in many ways human blood) have larger conductivity values, but are also thermal capacitors, able to retain heat.

Since the surface area can also affect heat conduction occurring orthogonal to a plane, Fourier's law is also written as:

$$\frac{\Delta Q}{\Delta t} = -kA\frac{\Delta T}{\Delta x} \ .$$

Where A is the cross-sectional surface area, ΔT is the temperature difference between the ends of a thickness, and Δx is the distance between the ends.

The thermal conductivities for various materials, including human tissue, are listed below.

Material	$k(W/mK)$
Human tissue (organs/muscle)	0.5
Human tissue (fat)	0.2
Human tissue (skin)	0.3
Blood	0.5
Bone	0.5 – 0.6
Water	0.62
Air	0.03
Glass	1.1
Aluminum	200
Copper	400

As can be seen from the above list, metals have high thermal conductivities, which is to be expected. Most human tissue has relatively low thermal conductivities, suggesting that conduction is not rapid within the human body. The skin has a lower conductivity than human tissue, suggesting that conduction into the body (as on a hot day) is also limited. It is interesting to note that air, like most gases, has a very low thermal conductivity. Thus, trapped air within clothing provides excellent resistance to thermal conduction in either direction – into or out of the body. On an extremely cold day, a parka with trapped air within a nylon covering provides excellent thermal resistance.

A three layer model can be utilized to examine heat transfer via thermal conduction as can be seen in Figure 3.2.

By applying Fourier's Law to each layer, and assuming that there is no accumulation of heat within any layer, the equations become:

$$Q1 = K1(T1 - T2)/DX1 \quad Q2 = K2(T2 - T3)/DX2 \quad Q3 = K3(T3 - T4)/DX3 \ .$$

Since there is no accumulation of heat, then $Q1 = Q2 = Q3 = Q$.
Rearranging the above equations,

$$T1 - T2 = QDX1/K1 \qquad T2 - T3 = QDX2/K2 \qquad T3 - T4 = QDX3/K3 \ .$$

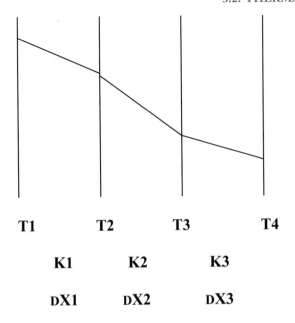

T1   T2   T3   T4

K1   K2   K3

DX1   DX2   DX3

Figure 3.2: A three layer model for steady state heat conduction.

Summing these equations results in:

$$T1 - T4 = Q/(DX1/K1 + DX2/K2 + DX3/K3)$$

or

$$Q = (T1 - T4)(DX1/K1 + DX2/K2 + DX3/K3) \, .$$

Since the thermal conductivities are in the denominator, any one term can affect the overall heat transfer rate. Thus, if air is one of the layers, its thermal conductivity is so low, that with $1/K$ as a factor, it would dominate the other layers. As a result, it is usually the layer with the lowest thermal conductivity that affects the overall heat transfer rate via conduction. This is why trapped air is such a good thermal insulator. Note that $T1 - T4$ will eventually equal out if there is no continual heat source, but the rate of transfer will be slower with air as one of the layers. If the body maintains a certain level of heat production, then a coat with trapped air will help to maintain skin and body temperature without a danger of hypothermia. A diagram taken from Ruch and Patton (1965) shows the temperature distribution inside and at the body surface in Figure 3.3.

Although the temperature nearest the skin is lower in a cold environment, the use of clothing with trapped air avoids the skin temperature becoming dangerously cold. For someone in an extremely cold environment, such as outer space, the use of an encapsulated suit is needed, as the exterior temperature is extremely low. In an earthbound environment, the issues associated with insulating

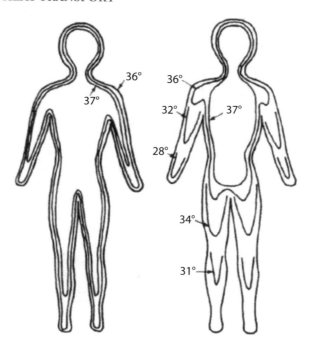

Figure 3.3: Temperature distribution in the human body, with the left figure depicting the temperature distribution in a warm environment and the right figure for a cold environment.

clothing do not address the added issue of heat loss (or gain) via respiration. We will soon see how respiration can be an important element in heat loss from the body.

3.3   THERMAL CONVECTION

Convection is heat transfer via bulk motion of a fluid, which can be a liquid or a gas. Examples of convective heat transfer include air conditioning coming from a compressor, blower and ductwork; wind chill; heat exchangers; and blood warming/cooling as part of heart/lung machines (also called cardiopulmonary bypass).

Convective heat loss can be *forced* or via natural (*unforced*) mechanisms. Forced convection is commonly due to a pressure gradient or other forcing function, which would move a fluid (with its associated heat) to another position or location. Thus, a forced air system consisting of an air conditioner, blower and vents is an example of a forced convection system. Another is wind chill, resulting from a pressure gradient moving air across a surface and removing heat from that surface. If an individual experiences wind chill, they would feel colder, as the wind chill would remove heat from the skin.

Natural or unforced convection is due to thermal gradients in a fluid, which results in fluid motion due to buoyancy differences. Thus, hot air rises as it is lighter than cold air. The movement of the hot air by differences in fluid density is a prime example of unforced convection. Normally, free convection is much smaller in magnitude than forced convection, as the movement of fluid via density differences is often far smaller than the movement of a fluid via pressure gradients. Hot air rises but at a rate far smaller than forced hot air from a heat pump through vents, or via wind chill from a strong wind.

Convective heat loss can generally be approximated by a convective heat loss coefficient used in the generalized equation

$$Q_c = h_c A \Delta T$$

where

$Q_c = $ convection heat transfer rate.

$h_c = $ convective heat transfer coefficient.

$A = $ area over which heat transfer occurs.

$\Delta T = $ temperature gradient between surface and environment.

Typically, the convective heat transfer coefficient is a function of fluid velocity for forced convective heat transfer. There are several versions of the equation relating the convective heat transfer coefficient to fluid velocity. One such example is

$$h_c = 5.6v^{0.67}$$

where v is in m/sec and h_c is in kcal/m^2 $-$ hr $-$ °C.

As the fluid velocity increases, so does the convection heat transfer, which increases the convective heat transfer rate. Thus, as the wind velocity increases, so does the wind chill, as an example of convection cooling. The convection heat transfer rate for free (natural) convection is normally a constant (@ 2 to 2.3. If one equates the free convective coefficient with the forced coefficient equation, then the air velocity required for the forced term to exceed the free term is $2.3 = 5.6v^{0.67}$ with $v^{0.67} = 2.3/5.6 = 0.4107$ and therefore $v = 0.263$ m/sec or $v = 0.588$ mph.

Thus, any wind speed over about a half mile per hour would exceed the free convective heat loss.

However, the more accurate method of computing the forced convective heat loss is to use the Reynolds number (indicative of laminar vs turbulent flow), the Prandtl number (relating viscous effects to thermal effects/conduction), and the Nusselt number (convective heat transfer to thermal effects/conduction). This allows for the specific calculation of the convection heat transfer rate for laminar or turbulent flow. As these two types of fluid flow are very different in terms of velocities and velocity profile, it is not surprising that the heat transfer associated with these two types of fluid flow are quite different.

In general, the Nusselt number (Nu) = $h_c D/k$ where D is the tube diameter or representative length, k is the thermal conductivity, and h_c is the convective heat transfer coefficient. However, the Nusselt number is related to the Reynolds number (Re) and the Prandtl number (Pr) by the following representative equations:

$$\text{Nu} = 0.322 \text{ Re}^{0.5} \text{ Pr}^{0.33} \quad \text{for laminar flow}$$

$$\text{Nu} = 0.023 \text{ Re}^{0.8} \text{ Pr}^{0.33} \quad \text{for turbulent flow.}$$

Where Re= $\rho V D/\upsilon$ and Pr = $c_p \upsilon/k$ with ρ = density, υ = viscosity and c_p = specific heat.

Since the Nusselt number is related to both the Reynolds number and the Prandtl number, as well as to the convective (forced) heat transfer coefficient, then by computing these three parameters, one can determine the convective heat transfer for the appropriate type of flow. Obviously, this is a more complex set of calculations than the simple form of $h_c = 5.6 \upsilon^{0.67}$, but it is more accurate since the simplistic form does not designate the type of fluid flow.

3.4   HEAT EXCHANGERS

Heat exchangers are commonly used in medical settings as blood heaters/coolers for open heart-lung machines or for patient heating/cooling for standard surgeries. The purpose of cooling the patient during surgery is to reduce the metabolic load, which lowers the need for larger blood flow rates or ventilation rates. The patient is then heated (or the blood heated) towards the end of surgery. A typical type of heat exchanger for blood is the double pipe heat exchanger, which consists of two concentric pipes with blood in one pipe and water in the other. By using cold water, the blood is cooled, while using warm water heats the blood. A double pipe heat exchanger combines convective heat transfer, as the fluids are moving, along with heat conduction through the pipe walls. A typical double pipe heat exchanger is shown in Figure 3.4. As the blood and the water can either be moving in the same direction (co-current flow) or in opposite directions (counter-current flow), there are two versions of the heat exchanger.

The physical device is shown in Figure 3.5.

The design equations for a double pipe heat exchanger combine Fourier's Law for heat conduction through the wall of the inner pipe together with the complex form of the convective heat transfer coefficients with the Reynolds, Prandtl, and Nusselt numbers.

This equation is $Q = UA\,\Delta T_{lm}$ where U is the overall heat transfer coefficient and is a function of the thermal conductivity and the convective heat transfer coefficient, A is the surface area of the pipe, and ΔT_{lm} is the effective temperature difference between the two fluids from one end of the exchanger to the other. This term is known as the log mean of the temperature gradient. Since the surface area of the pipe could be the inner or outer surface, then the overall heat transfer coefficient is synched with the surface area as U_o and A_o or U_i and A_i. The temperature gradient is unaffected by inner or outer diameter, but it is affected by whether the flow is co-current or counter-current. In general, counter-current flow is more efficient for overall heat transfer.

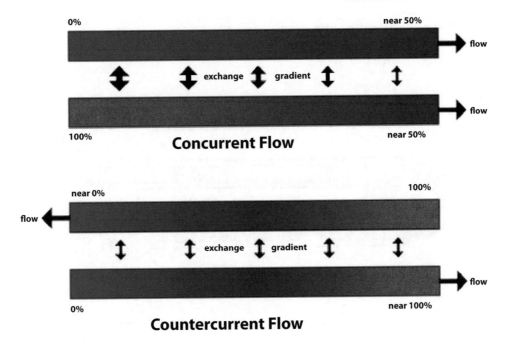

Figure 3.4: Double pipe heat exchangers for co- and counter-current flow (Wikipedia).

3.5 THERMAL RADIATION

Thermal radiation is *electromagnetic radiation* emitted from a material which is due to the *heat* of the material, the characteristics of which depend on its *temperature*. An example of thermal radiation is the *infrared* radiation emitted by a common household *radiator* or *electric heater*. A person near a raging bonfire will feel the radiated heat of the fire, even if the surrounding air is very cold. Thermal radiation is generated when heat from the movement of *charges* in the material (*electrons* and *protons* in common forms of *matter*) is converted to electromagnetic radiation. Sunshine, or *solar radiation*, is thermal radiation from the extremely hot gasses of the *Sun*, and this radiation heats the *Earth*. The Earth also emits thermal radiation but at a much lower intensity because it is cooler. The balance between heating by incoming solar thermal radiation and cooling by the Earth's outgoing thermal radiation is the primary process that determines the Earth's overall temperature. As such, radiation is the only form of heat transfer that does not require a material to transmit the heat. Radiative heat is transferred from surface to surface, with little heat absorbed between surfaces. However, the surfaces, once heated, can release the heat via conduction or convection to the surroundings.

Thermal radiation is conducted via electromagnetic waves. As such, this form of heat transfer is not only a function of the temperature difference between the two surfaces, but it is also the frequency range of the emitted and received energy. As an example, sunlight is composed of the

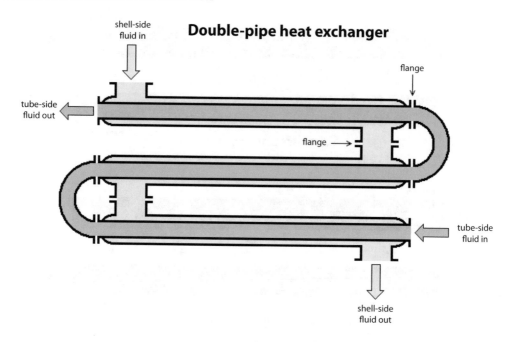

Figure 3.5: Double pipe heat exchanger design (Wikipedia).

visible light spectrum, as well as infrared energy and ultraviolet energy. Figure 3.6 depicts the effects of temperature and wavelength of the thermal energy on the heat transfer rate.

When radiant energy reaches a surface, the energy can be absorbed, transmitted (through) or reflected (or any combination). The sum of these three effects equals the total energy transmitted, and the parameters which describe these three phenomena are given by:

$$\alpha + \rho + \tau = 1$$

where

α represents spectral absorption factor,

ρ represents the spectral reflection factor, and

τ represents the spectral transmission factor.

The radiative heat transfer rate is given by the Stefan–Boltzmann law:

$$Q_r = \sigma A T^4$$

where σ is the Boltzmann constant and A is the surface area of the radiating source. The temperature is in an absolute scale (degrees Kelvin, corresponding to degrees C or degrees Rankin corresponding to degrees F).

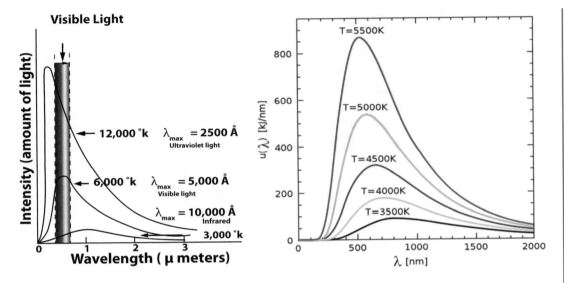

Figure 3.6: Peak wavelength and total radiated amount vary with temperature. Although the plots shows relatively high temperatures, the same relationships hold true for any temperature down to absolute zero. Visible light is between 380 to 750 nm.

To predict the exact amount of radiative heat transfer between two surfaces, the above equation is expanded as follows:

$$Q_r = \sigma F_1 A_1 \left(T_1^4 - T_2^4\right)$$

where F is the facing factor that represents the amount of the emitting surface (1) facing the receiving surface (2) with the surface area A representing surface 1. Correspondingly, this equation can use F_2 and A_2 to represent the facing factor for the receiving surface towards the emitting surface with the surface area of 2. Boltzmann's constant and the temperature gradient are unchanged for either form of the equation. The facing factor can be approximated as a disk of radius R if the distance between the two surfaces is large, such as the earth to the sun. For shorter distances, the facing factor is a complex interaction between the angles of the two surfaces that face each other.

As can be seen, thermal radiation is affected by the frequency of the emitted energy. This is why sunscreen ointments have ultraviolet protection, as this type of energy can be damaging to skin. In addition, it is common to feel warmer on the sunny side of the street as opposed to the shady side, given the radiative heat transfer. Radiation can be a significant source of heat as compared to the other forms (conduction and convection) as radiation is composed primarily of sunlight and the respective heating of the earth.

3.6   GREENHOUSE EFFECT

Emitted radiation from the sun consists of a spectrum of electromagnetic radiation of varying frequencies and wavelengths. The sun's surface temperature is approximately 10,000 degrees C. When this energy hits the earth's surface, it can be absorbed, reflected, or transmitted through. The amount of each of these factors depends on the surface temperature of the earth and the exact material (water, soil, or man-made materials). However, the reflectivity, absorptivity or transmissivity are frequency dependent. In addition, the frequency spectrum of the incoming thermal energy is shifted due to the different surface temperature of the earth as compared to the sun.

As the energy reached the earth through its atmosphere, the transmissivity was high (allowing it to pass through the atmosphere) and the reflectivity and absorptivity were low. However, as the energy hit the earth's surface, the new temperature shifted the spectrum such that the new reflectivity was high and the transmissivity was low. Thus, when some of the thermal energy was reflected back from the earth's surface towards the atmosphere, it did not transmit through towards outer space, but instead was reflected back towards the earth's surface. At each reflection back, more energy was absorbed into the earth's surface, heating the earth. Normal conduction and convection would heat the surrounding air and buildings, resulting in warm temperatures that allows us to live on this planet. This phenomenon is known as the *greenhouse effect*.

This is shown in Figure 3.7.

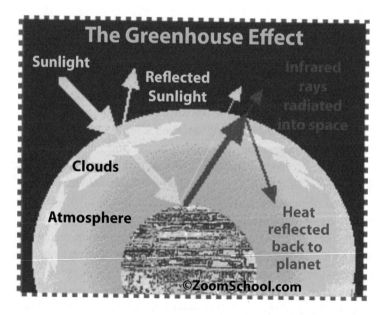

Figure 3.7: Depiction of the greenhouse effect resulting in thermal radiation being reflected back towards the earth's surface (Wikipedia).

You probably know the greenhouse effect first hand if you leave your car unattended on a sunny day. The thermal radiation is transmitted through the windshield and hits the seats and dashboard. The frequency shift occurs and the reflected energy is then reflected back to the windshield, but now at a high reflectivity and low transmissivity. As such, the inside of the car heats up, often to 20-30 degrees higher than the ambient air. A sunscreen on the window will reduce the amount of radiation entering the car and keep it cooler.

Commercial greenhouses purposefully trap the heat via the greenhouse effect and use some limited screening to control the amount of trapped heat. The increased heat allows the plants to remain at a relatively warm temperature and avoid the changes in temperature with changing air currents and air temperature. By keeping the doors to the greenhouse closed, the temperature can be somewhat maintained overnight. A commercial greenhouse is shown in Figure 3.8.

Figure 3.8: Commercial greenhouses.

3.7 HEAT LOSS VIA RESPIRATION

When air is inspired into the lungs, heat and water vapor are transferred to the air by convection and evaporation from the surface lining the respiratory tract. By the time the air has reached the deepest parts of the lungs, the air is at deep body temperature (37 degrees C) and saturated with water vapor (47 mm Hg partial pressure). As the air moves outward through the respiratory tract during expiration, some heat is transferred back to the body and some water is condensed. However, the inspired air still contains significantly more heat and water than the inspired air.

Respiration results in a latent heat loss and a sensible heat loss. The latent heat loss is based upon the latent heat of vaporization of water by the following equation:

$$Q_{el} = (dm_a/dt)(Y_o - Y_i) \quad \lambda$$

where

(dm_a/dt) is the kilograms of air breathed in and out/hr.

$(Y_o - Y_i)$ is the difference in the expired and inspired air water content.

λ is the latent heat of vaporization of water at the expired air temperature.

The pulmonary ventilation rate (dm_a/dt) is primarily a function of the body metabolic rate via the following relationship:

$$(dm_a/dt) = 0.006M$$

where M is the metabolic rate in Kcal per hour.

It is somewhat rare to breath in bone dry air, so the inspired air temperature and water content can affect the expired air from the lungs. Fanger et al. (1968) found that there is a relationship between the inspired water content of air and the expired water content:

$$Y_o = 0.029 + 0.20Y_i$$

for normal conditions. Thus, if the entering air is very dry ($Y_i = 0$), the expired air would be less humid than if the inspired air were humid. McCutchan and Taylor (1951) determined that the temperature of the expired air is dependent on the inspired air in the following fashion:

$$T_o = 32.6 + 0.066\, T_i + 32\, Y_i\ .$$

Although the relationship between T_o to T_i is expected, the inspired air humidity also plays a role. The air, upon being expired, is somewhat cooler than it would be, had the inspired air been humid.

Besides the evaporative, latent heat loss associated with respiration, there is also a sensible heat loss (when $T_o > T_i$) or heat gain (when $T_o < T_i$) which is described by:

$$Q_{sl} = (dm_a/dt)C_p(T_o - T_i)$$

where C_p is the specific heat.

This is the heat loss due to heating up the dry air component of the inspired air. An example of calculations for heat loss is as follows:

Assume that 6 liters/min of bone dry air at 20 degrees C is inspired (12 breaths per minute at 500 ml tidal volume per breath = 6 liters per minute); also that the expired air is fully saturated with water vapor and is at 37 degrees C. The physical properties are:

$C_{p,air}$ at 20 degrees C $= 0.25$ cal/g-degree C

λ at 37 degrees C $= 577$ cal/g

Vapor pressure of water at 37 degrees C $= 47$ mm Hg.

The dry air mass flow rate in grams per minute can be calculated via the ideal gas law:

$$(dm_a/dt) = (6 \text{ liters/min})(1 \text{ mol}/22.4 \text{ liters})(273 \text{ degrees C}/293 \text{ degrees C}) (28.9 \text{ g/mol})$$
$$= 7.2 \text{ g/min of dry air.}$$

The amount of water vapor in the expired air is

$$dm_w/dt = (7.2/28.9)(47/760 - 47)(18) = 0.295 \text{ g/min.}$$

Note that the true ration of the partial pressure of water vapor in the lungs is not 47/760, but rather 47/760-47, as the water vapor is attached to the dry gas, which is the remainder of the partial pressure (713). The molecular weight of water is 18. It is necessary to compute the moles of water vapor when multiplying by the partial pressure ration, as the partial pressures are related to the moles, rather than the mass. Multiplying by the molecular weight returns the calculation back to units of mass.

The latent heat loss is:

$$Q_{el} = 0.295(577) = 170 \text{ cal/min.}$$

The sensible heat loss is:

$$Q_{sl} = 7.2(0.25)(37 - 20) = 30.4 \text{ cal/min .}$$

Note that the latent heat loss is far more significant than the sensible heat loss. The total heat loss of approximately 200 cal/min equals 12.0 kcal/hr, which is 17% of the basal metabolic rate of the body – an appreciable quantity. During exertion, one's respiration increases, which provides even greater heat loss by this mode. The overall rise in the metabolic heat production rate is also greater, such that an additional heat loss mode is necessary, such as sweating. Evaporation of sweat from the body produces the same latent heat loss as was evident via respiration. However, if one sweats profusely, such that sweat rolls off the body, then dehydration may occur, which can be quite dangerous.

An interesting note is that the panting of dogs is the primary mechanism of heat rejection, since dogs cannot sweat. By interchanging air with the dead space on a rapid basis, rather than via deep breathing, the dogs avoid hyperventilation, yet they produce heating of inspired air as well as humidification. Thus, there is latent and sensible heat loss. Many dogs have considerable fur coats, which provide insulation against heat gain or loss via conduction. Thus, if the dog were to get overheated, the solution is to hose the dog down, which would reduce the trapped air insulation via the fur. The same effect occurs when skiing in a cold environment. A ski parka has trapped air which serves as an insulator. However, were the parka to become wet, the trapped air would be mixed with water, thus reducing its effectiveness as an insulator.

3.8   HEAT LOSS INSIDE THE BODY

Many of the mechanisms for heat transfer which have been discussed refer to heat loss from the body to the environment. However, heat must be transferred from the body core to the skin or lungs. The two mechanisms available for such transfer are via conduction and via convection. Conduction is a slow process. Convection is accomplished via blood flow. As blood reaches the body core, it absorbs heat. Blood then is channeled towards the periphery where heat is released. Ruch and Patton (1965) discuss the mechanisms by which the blood circulation affects internal heat distribution in three ways:

1. It minimizes temperature differences within the body; tissues having high metabolic rates (e.g., liver) are more highly perfused, and they are thus kept at nearly the same temperature as less active tissues. Cooler tissues are warmed by blood coming from the active organs.

2. It controls the effective body insulation in the skin region; warm blood is increased to the skin via vasodilation when the body wishes to reject heat; blood is bypassed from arteries to veins via deeper channels through vasoconstriction when conservation of body heat is vital.

3. Countercurrent heat exchange between major arteries and veins often occurs to a significant extent; if heat conservation is necessary, arterial blood flowing along the body's extremities is pre-cooled by loss of heat to adjacent venous streams.

A model of heat transfer from the core towards the skin can be split into two sections – from the core to a muscle region and from the muscle region to the skin (Cooney, D., 1976). The heat transfer includes both a conductive term using Fourier's law and a convective term using a forced convective form as follows:

$$Q = k_{cm} A \frac{(T_c - T_m)}{\Delta Z_{cm}} + (dm_b/dt) C_{pB} (T_c - T_m)$$

$$\text{conduction} \qquad\qquad \text{convection}$$

$$Q = k_{ms} A \frac{(T_m - T_s)}{\Delta Z_{ms}} + (dm_b/dt) C_{pB} (T_m - T_s) \,.$$

Since the heat transfer rate is assumed to be the same for both zones and the convection term is likely far greater than the conduction term, then an overall heat transfer rate from core to skin is of the form:

$$Q = (dm_b/dt) C_{pB} (T_c - T_s) \,.$$

This form does not account for any shunting of blood to allow for heat gain or loss via vasoconstriction or vasodilation.

3.9   HEAT LOSS IN EXTREME ENVIRONMENTS

One example of heat loss (or gain) in an extreme environment is in a fire and is experienced by firemen. As such, fire protection clothing and gear must protect against excessive heat conduction, and therefore must have considerable insulating effect. It must also protect against radiation, as an open flame produces significant radiant energy, as was shown in Figure 3.1. As water is used at high flow rates, the fireman's suit must also be water resistant. A typical fireman's clothing, called *turnout gear*, is shown in Figure 3.9.

1 Nomex *hood*.

2 Cotton t-shirt with department logo small on chest and large on back.

3 Suspenders with retroflective striping, connecting to the pants at eight points.

4 Insulated pants with retroflective striping. They are held closed with velcro and spring hooks. They are reenforced with leather at the knees and bottoms, and they have two large side pockets and one smaller rear pocket.

5 Steel toed insulated rubber *boots*, with Vibram soles. These boots have handles at their tops to help pull them on, and they come up to just below the knees.

6 *Helmet*, with goggles and department logo. These helmets have a velcro/buckle chinstrap, adjustable headband, and a protective cloth flap that hangs over the collar, further protecting the neck and preventing embers from falling down the collar.

7 Goggles, attached to helmet. Used for eye protection when not wearing breathing apparatus.

8 Radio with clip-on microphone. These radios are waterproof and inherently safe (non-sparking) for use in explosive atmospheres if necessary.

9 D-ring Carabiner, used to clip additional equipment to the coat (not standard issue).

10 *Flashlight*. Department jackets have custom loops to hold the flashlights at center-chest.

11 Insulated leather *gloves*.

12 Insulated jacket with retroflective striping. Oversized pockets hold radio, gloves, a hose strap, etc. Like the pants, it is held closed with velcro and spring hooks.

13 Air-line and pressure gauge. On this particular brand of SCBA, there are two air gauges: one at the bottom of the tank in back (for checking the pressure when the tank isn't being worn) and one in front.

14 *SCBA* harness, comprised of shoulder and waist straps.

15 *PASS device*. Current issue is an integrated PASS/SCBA, which activates automatically when the air supply is turned on.

16 *Name label* on back of helmet.

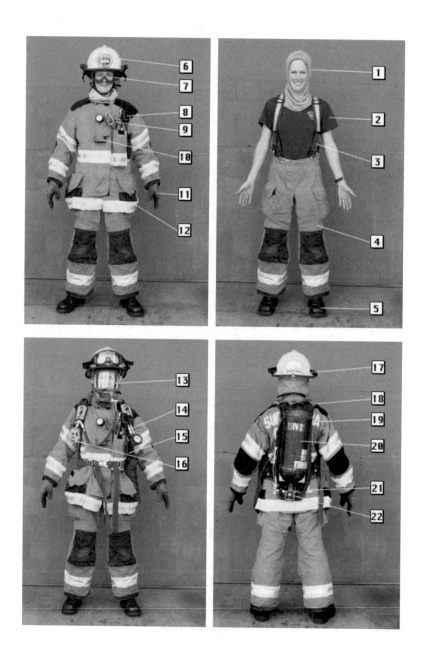

Figure 3.9: Typical fireman's turnout gear (Santa Clara Fire Dept.).

17 SCBA shoulder straps.

18 Department identification.

19 *Air tank* bottle and backpack frame. The bottle is quick-swappable, because at a working fire a firefighter often goes through several bottles.

20 Regulator and main supply valve.

21 *Name label* (under tank).

The coat and pants are insulated and reenforced. They are made out of a fire-resistant fabric called PBI (other materials like Nomex are also used). They have reflective stripes to make them reflect when a light is pointed at them, so that they can be better seen in the dark, as well as glow-in-the-dark patches. They also have the firefighter's name and department printed on the back, like a football player, to help identify them, since when everyone is suited up and wearing masks, it is difficult to tell who is who. They are also equipped with several large pockets for holding gloves, tools, radios, etc. Rubber or leather waterproof steel-toed boots protect the firefighter's feet. The rubber boots are usually stored within the 'turned out' pants so that they can be quickly donned, hence the term "turnouts." A fire-retardant hood covers the firefighter's head and neck, protecting ears and other parts that would be exposed under a helmet. When properly worn, no part of the firefighter's skin is exposed or unprotected.

Helmets are color coded, so that the wearer can be quickly identified at a fire scene. For many fire departments, the following color codes are used as seen in Figure 3.10.

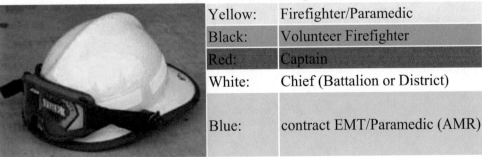

Helmet Color Coding	
Yellow:	Firefighter/Paramedic
Black:	Volunteer Firefighter
Red:	Captain
White:	Chief (Battalion or District)
Blue:	contract EMT/Paramedic (AMR)

Figure 3.10: Fireman protective helmet with color coding to determine function (Santa Clara Fire Dept.).

Bibliography

Cooney, D.O., (1976) *Biomedical Engineering Principles: An Introduction to Fluid, Heat, and Mass Transport Processes*, Marcel Dekker Publisher, New York, 1976. 56

Fanger, P.O., McNall, P.E., and Nevins, R.G., (1968) Predicted and measured heat losses and thermal comfort conditions for human beings, *Symposium on Thermal Problems in Biotechnology*, ASME, New York, 1968. 54

McCutchan, J.W. and Taylor, C.L., (1951) Respiratory heat exchange with varying temperatures and humidity of inspired air, *J. Applied Physiology*, vol. 4, p. 121, 1951. 54

Ruch, T.C. and Patton, H.D., (1965) *Physiology and Biophysics*, 19th edition, Saunders Publishers, Philadelphia, 1965. 45, 56

Santa Clara, CA Fire Department. 58, 59

Wikipedia. 7, 12, 14, 29, 42, 49, 50, 52

Author's Biography

GERALD E. MILLER

Dr. Miller is the Department Head of Biomedical Engineering at Virginia Commonwealth University as well as a professor of physiology, professor of cardiology, and a professor of physical medicine & rehabilitation. The Biomedical Engineering Department maintains graduate degrees leading to the M.S. and Ph.D. with 9 primary, 60 affiliate faculty, and 65 graduate students as well as Virginia's first undergraduate BME program with an enrollment of 250. Dr. Miller received the B.S. in aerospace engineering from the Pennsylvania State University in 1971, the M.S. in bioengineering from the Pennsylvania State University in 1975, and the Ph.D. in bioengineering from the Pennsylvania State University in 1978. He is a member of Phi Kappa Phi as well as a Fellow of both ASME and AIMBE. His research activities include rehabilitation engineering, physiological fluid mechanics, artificial internal organs, epilepsy genesis, and the use of physiological signals in the control of mechanical systems. He directs a Whitaker Foundation center in man-machine interfaces and co-directs the Bioengineering and Bioinformatics Summer Institute, a joint NSF-NIH program.